Harvard Studies in Business History, 42

Published with the support of the Harvard Business School
Edited by Thomas K. McCraw
Isidor Straus Professor of Business History
Graduate School of Business Administration
George F. Baker Foundation
Harvard University

News over the Wires

The Telegraph and the Flow of Public Information in America, 1844–1897

Menahem Blondheim

Harvard University Press
Cambridge, Massachusetts
London, England
1994

Printed in the United States of America

This book is printed on acid-free paper, and its binding materials have been chosen for strength and durability

Library of Congress Cataloging-in-Publication Data

Blondheim, Menahem, 1954–
News over the wires : the telegraph and the flow of public information in America, 1844–1897 / Menahem Blondheim.
p. cm. — (Harvard studies in business history ; 42)
Includes bibliographical references and index.
ISBN 0-674-62212-X
1. Journalism—United States—History—19th century. 2. Newspaper publishing—United States—History—19th century. 3. Telegraph—United States—History—19th century. I. Title. II. Series.
PN4864.B76 1994
071'.3'09034—dc20 93-28731
CIP

In loving memory of my mother, Syril Appleton Blondheim

Editor's Introduction

Without modern communication, life as we know it would be impossible. The "wired nation" of telephone and cable television lines, the international financial transactions made possible by satellite transmissions, the overnight creation of celebrities—all of these, for good or ill, depend on modern communication.

The United States first became a wired nation with the invention of the telegraph in the 1840s and its development over the next three decades. The crucial role of the new technology introduced by Samuel F. B. Morse in the development of the railroad industry is fully documented in American historiography. Less well known is the telegraph's equally important part in the maturation of the newspaper industry and the formulation of national news—processes that centered around the growth of the New York Associated Press and the Western Associated Press, their eventual merger, and their protean relationship with Western Union, the national telegraph monopoly.

In this path-breaking book, Menahem Blondheim meticulously delineates the evolution of the Associated Press. Weaving business history with cultural history, he shows how a small number of innovators recognized the time value of information about the price movements of American commodities in European markets. These innovators exploited the new telegraph technology to gain business advantages over their rivals. They made money both for themselves and for the newspapers that purchased their services.

Blondheim explains how the growth of the penny press in the 1830s set the stage for newspaper exploitation of the telegraph in the 1840s, and how these two developments in turn facilitated a "nationalizing" of the news. From a focus on local concerns, newspapers began to pay more and more attention to national and international affairs, with the Associated Press acting as a "giant funnel," channeling news from all over the country into its own offices. There the information was quickly winnowed, standardized, and sent back out to a national audience of newspaper readers. By the 1870s, the AP was the telegraph's biggest customer, accounting for 10 percent of Western Union's revenues. By the 1880s, more than 80 percent of the copy contained in western American newspapers consisted of AP dispatches.

This standardization of information raised some troubling questions for a free society. Was it legitimate for the federal government to take control of the national telegraph network and ally itself with the Associated Press, as it did during the Civil War? In such a situation, was not censorship inevitable? Even in peacetime, did not the standardization of data by anonymous "reporters" (a term originated by the AP) impart a spurious objectivity to the information now defined as "news"? Was any real objectivity possible? In one of his best chapters, Blondheim shows that most AP coverage during the late nineteenth century carried a slant favorable to the Republican party, and that it may have affected national elections from the 1870s to the 1890s, when his history ends.

Blondheim's treatment of these issues is no dry recounting of bureaucratic empire-building by nameless functionaries. Rather, he introduces such lively characters as Daniel Craig, the key Associated Press entrepreneur and the hero of the book; Craig's archrival "Fog" Smith, an appealing scoundrel who could have been invented by Mark Twain; and the wily Jay Gould, corporate raider, sometime owner of Western Union, and habitual fisher in troubled waters.

In his broader interpretive scheme, Menahem Blondheim advances the provocative thesis that a "news monopoly" resulted from the workings of the Associated Press. Even readers who do not agree with every detail of the author's viewpoint will benefit greatly from his book. No other student of the subject has done nearly the amount of primary research here accomplished by Blondheim, and the historiography of American business and American culture is much the richer for it.

—Thomas K. McCraw

Contents

News over the Wires

Introduction

"He actually wields more power for good or evil, more power to raise or depress the fortune of the country, more power to make men and women rich and famous or poor and infamous, than any man in America." Thus was William Henry Smith, joint general manager of both the New York and the Western Associated Press, introduced to the readers of the *San Francisco Post* during a visit to the West Coast in 1885. The account enumerated Smith's powers: he could bring "railroad presidents to . . . their knees" and mighty politicians "to grovel for his favor," or even be kissed by "those most imperious of all human beings," the country's great prima donnas.

The *San Francisco Post* reporter explained the source of Smith's considerable power quite simply: "William Henry has got the attentive ear of more men and women than any other man in the country." When Smith impressed a point on the minds of 13,000,000 newspaper readers, "it [came] pretty near being crystallized into the history of the country."[1] A less deferential account, using more modern terminology, would claim that the power of the Associated Press and its officers was based on a monopoly of knowledge. Communication systems, by nature, allow the accumulation of information in certain strategic junctures. By 1885 the Associated Press commanded these key positions in the country's system of news flow. It had exclusive possession of the confluence of news in America—local, national, and international—before it reached anyone else. By virtue of this monopoly it had the attentive ear of that important segment of the American public that wanted to know the news.

"Let a man only tell you his story every morning and evening," proposed Edmund Burke, "and at the end of a twelvemonth he will have become your master."[2] From a broader perspective, the media scholar

James Carey suggested that a monopoly of knowledge in its cultural sense implied the power "to determine the entire world view of a people: to produce . . . an official view of reality which can constrain and control human action."[3] The Associated Press had that potential power to "create and define the world" for the American public once it secured a monopoly in the national distribution of news. In the following chapters I will examine how and why this monopoly was created, and to what effect.

The concept of a monopoly of knowledge, and the preliminary investigation into its formation and significance, are widely identified with the work of Harold Adams Innis, a student of Canadian economic history turned communication theorist.[4] His application of the dynamics of oligopoly to the supply of, and demand for, information suggests the benefits of investigating the sociology of knowledge and its politics from a business perspective. Innis's approach to the study of communications was historical, and its scope ecumenical. He described the emergence of monopolies of knowledge such as those held by Egyptian priests invested with exclusive knowledge of a sacred language, by Ptolemy's magnificent library in Alexandria, where the world's knowledge was assembled, and by secluded medieval monasteries, in which parchment scrolls were studied and copied so that the same could be done in other secluded monasteries.

Such notions of exclusiveness and restriction were refreshingly foreign to the United States in the first half of the nineteenth century. In America, procuring most types of news appeared to be free to all and beyond monopoly. An earthquake or the arrival of a steamer had many witnesses. Sermons, court verdicts, debates, and laws achieved wide audiences. Moreover, the gathering of news was ventured upon in most cases precisely for the purpose of diffusing it widely rather than retaining it exclusively. And indeed, in early-nineteenth-century America, the distribution of news appeared to be anything but vested in a monopoly. Visitors from Western Europe were uniformly impressed by the unparalleled profusion of newspapers among the masses. The keenest observers saw a relation between the extent of the diffusion of knowledge by the press and the structure of power and control in American society. There was a "necessary connection between public association and newspapers," noted Alexis de Tocqueville. "Newspapers make associations and associations make newspapers." He found that in the United States there was both "the greatest number of associations and of newspapers," and he concluded that "the power of the newspaper press must therefore increase as the social conditions of men become more equal."[5]

Nevertheless, the American situation demonstrated the full power of

Innis's concept of a monopoly of knowledge. Scrutinizing news gathering and news flow in early-nineteenth-century America, Innis pointed out a significant bias in the availability and control of news: a bias in the dimension of space. Important news items were characteristically available first in the leading mercantile and political cities. From these geographic centers they radiated to the periphery in the form of newspaper articles copied from one journal to another. Although the news was ultimately widely diffused, and via many newspapers, the press of the geographic centers monopolized news sources and enjoyed the advantages of early possession of the news. The timeliness of newspaper accounts declined in direct relation to distance from the center.

Political authorities came to realize that by controlling organs at the center, they could dominate knowledge at the periphery. Once established, such organs became prime sources of important political information, and in turn the prominence of the press of the political and economic centers was further magnified. According to Innis, it was, paradoxically, the abundant flow of news and the wide diffusion of the newspaper that enhanced central control.[6] The "untrammeled freedom to twist and misrepresent the news" was thus enjoyed in essence by only a limited number of papers that were the first-hand source for the country's news.[7] "Freedom of the press," Innis maintained, "has been regarded as a great bulwark of our civilization [but in fact] it has become the great bulwark of monopolies of the press."[8]

The telegraph, which first became available for public use in the United States in 1844, had the potential of making information available simultaneously throughout the country; it could become a great equalizer. Indeed, Innis considered the diffusion of the telegraph a major turning point, bringing decisive social and political change.[9] He argued that "the telegraph weakened the system of political control through the post office and the newspaper exchange. The monopoly over news was destroyed and the regional daily press escaped from the dominance of the political and the metropolitan press."[10] Moreover, when the monopoly of geographical centers over knowledge was abolished, central authority itself was threatened. By "annihilating time and space," the telegraph destroyed "the power of party based on . . . state capitals" to the extent that "power shifted from Washington and issues were no longer settled in Congress."[11] But where did power shift to? Had the telegraph, by abolishing the advantages of geographic centers, finally destroyed the power to monopolize and control information?

Marshall McLuhan, a self-proclaimed follower of Innis, thought so. What made the telegraph especially significant in McLuhan's view was its potential to connect individuals across space, both in fact and in consciousness. "Electric media," asserted McLuhan, "abolish the spatial dimension . . . By electricity, we everywhere resume person-to-person relations as if on the smallest village scale. It is a relation in depth, and without delegation of functions or powers."[12] McLuhan's vision of the electronic media's creating a global village was predated by more than a century (and dozens of bloody wars) by Samuel Finley Breese Morse's "one neighborhood of the whole country." Morse's expectation that the telegraph would be largely employed for social intercourse was echoed in McLuhan's perception of the device as "the social hormone."[13] Both Innis and McLuhan held that oral communications, being imminently inclusive and interactive, were the ideal system for a libertarian society. With the telegraph making instantaneous two-way communications possible, McLuhan believed this ideal was reached: "Dialogue supersedes the lecture."[14]

McLuhan and Innis presented separate analyses of the impact of the telegraph in relation to its technological capabilities. The former emphasized its interactive capacity as the key departure; the latter, its power to diminish signaling time across space. Both these capabilities tended to promote a free and equal flow of information and to prevent monopoly of knowledge. Yet telegraphic technology also worked in the opposite direction, toward monopolistic control of knowledge. The early telegraphs transmitted information sequentially; each message had to be sent separately and in turn. Whereas the mailbag could carry numerous communications simultaneously, the telegraph transmitted only a single copy of one message at a time. The telegraph line thus promoted equality of knowledge across space, at the price of a monopoly for certain messages in time.

Then, too, the telegraph promoted monopoly because of its channel-like nature. Turnpikes and canals also restricted communications to a channel, but unlike them, and very much like the railroad, the telegraph was a single unit that was both the road and the vehicle. Control of telegraph lines implied control of the message traffic flowing through them.[15] When information was moved along canals and roads, numerous vehicles could compete over a variety of channels; but there was only a single conduit for instantaneous communications, and it could easily be monopolized. Thus, the telegraph had some attributes that tended to abolish monopoly over knowledge, and others that facilitated it. Promoters' interests and users' requirements determined which of these attributes would prevail and how its functions would be shaped. Only the interaction of technology and environment determined the ultimate impact the telegraph had on society. To study that impact, I have investigated the emergence, development, and significance of the institution of telegraphic news gathering and distribution in the United

States. In particular, I have focused on the major American news wire service—the Associated Press.[16]

A business history perspective, enlightened by insights and ideas developed in communication theory, guided my attention to particular aspects of wire-news history. Given this orientation, and in an attempt to apply the dynamics of oligopoly to supply and demand for information, I focused on a spectrum of critical episodes. These uncanny struggles and lively skirmishes, minor catastrophes and their patchy solutions provide in turn an indispensable perspective on the more general forces—economic, social, and political—that shaped the interaction of news and wires. After all, the significant institutional innovations in the course of wire service development were reached by actors trying and erring in the process of shaping a new information ecology, within a rapidly changing society, polity, and business environment.

Accordingly, I gained my view of the wire services, in part, by looking over the shoulders of nineteenth-century editors at work managing their newspapers, by glimpsing perplexed telegraph entrepreneurs struggling to diffuse their novel technology, and by making the acquaintance of the men at the helm of the wire service. Of the accomplishments of these men, the *res gestae* of Daniel H. Craig, the real hero behind the emergence of the wire services as a national institution, were the most important. A prototype of the professional manager, Craig was one of the most effective but least known, most colorful but least appreciated, most imaginative and most controversial of the country's great system builders. Close-ups of the adventures of William Henry Smith, juggling political and economic interests, maneuvering between regions and among newspaper editors and business leaders, were no less revealing. They provided insight into the sensitive interests the wire service could and did effect, and they demonstrated the central position it had acquired by the late nineteenth century. It is through interweaving these strands—the events, the lives, and the times—that the large tapestry emerges, presenting an important agent in the modernization of nineteenth-century America.

A study of this kind depends, very heavily, on a solid body of knowledge of American journalism, of telegraphy, and, particularly, of the Associated Press and other news wire arrangements. Much of this knowledge, however, had to be gleaned from primary sources, many of them not previously studied in any depth; for although useful scholarly work on the history of American journalism abounds, and much is also available on telegraphy, historians have given surprisingly little attention to the development of the nineteenth-century national news wire services. The pioneering effort of Victor Rosewater, *History of Coopera-*

tive News-Gathering in the United States, published in 1930, has only recently been supplemented by Richard Schwarzlose's masterly book *The Nation's Newsbrokers.*[17]

A variety of reasons can account for the relative neglect of this field by American historians. The sources are extensive but widely scattered. More important, the Associated Press was a relatively small organization: it did not own real assets, its employees were thinly scattered, and their number was no greater than that of any one of the leading metropolitan newspapers. Clearly, historians of American business found larger and richer organizations to study. Moreover, the history of telegraphic news has had very low visibility; its development was what recent historians call a "hidden history." The Associated Press had a strong preference for operating backstage, leaving the limelight to the newspapers it served and to their editors. Or perhaps editors did their best to conceal their dependence on the wire service from their readers. For example, the *New York Times* indexes for the 1850's include only seven references to the Associated Press, and AP datelines in the American press were rare before the 1880's.

This low visibility was compounded by the unique position of the wire service in the field of business arrangements. It operated on the borderline between, and on the margins of, the press and the telegraph. Even when focusing on either of these larger industries, most historians failed to perceive the autonomous institutional nature of the news and wires nexus. The significance of its unique organizational structure, managerial practices, and revolutionary business strategies was dimmed when viewed as an aspect of, or an auxiliary to, either journalism or telegraphy.

But, however small and invisible the organization, its effects were widespread and very real. For by the early 1850's at least two columns of Associated Press news appeared daily in nearly every major American newspaper. And contemporary readers purportedly considered them the most compelling feature of the newspaper. Historians of journalism, however, have not found these columns interesting and gave them scant attention. The reports were uniform in all newspapers that had wire service privileges, and, from a literary perspective, they were quite boring. Studying the columns of nineteenth-century Associated Press telegraphic news is a chore only as interesting as verbatim accounts of filibusters in Congress, assorted stock quotations, and pages from the ledger of a compulsive clerk in a seaport can be. Wire news was merely the nuts and bolts of American history in daily installments. It was in other columns of the nineteenth-century newspaper that historians found some of the most vivid and picturesque specimens of journalism ever produced. It is not surprising, then, that Associated Press wire reports were seldom if ever included in anthologies of nineteenth-century journalism, monographs on newspapers, or biographies of editors.

Yet I believe that by focusing on telegraphic news in general, and the Associated Press in particular, some of the major problems in the history of American journalism can be addressed, not least among them the issues of timeliness in, and monopoly of, news. Moreover, the emergence of the Associated Press and, more generally, of telegraphic news, had an important bearing on national development in nineteenth-century America. The Associated Press was probably the first private-sector monopoly to operate on a national scale in the United States; its business was knowledge. The Associated Press, by securing for itself the position of a national news monopoly early in the second half of the nineteenth century, represented one of the most powerful centripetal forces shaping American society in the modern era.

PART ONE

The Emergence of the News Wire Services

1

"The News Newer": Discovering Timeliness in News

"The greater the speed with which intelligence can be transmitted from point to point, the greater is the benefit derived to the whole community."[1] This was the rationale Samuel Finley Breese Morse used in urging that his electro-magnetic telegraph be adopted to speed the flow of information in America. The evaluation of speed in terms of social good is an issue open to moral judgment; yet how accelerated communication affects society is an important but neglected historical question.[2]

Journalists, managing the main medium for diffusing public information in America, had been preoccupied with speed long before the invention of the revolutionary device for the "annihilation of time and space." In fact, at the time the telegraph was introduced, they were expending enormous efforts to abolish as much as possible the detrimental effect time and space had on their business. Clearly, the full impact the telegraph had on the press can only be appreciated by understanding the significance of the speed of news to journalism.

The telegraph, then, did not inaugurate the movement to accelerate the flow of American news, nor were the savings in time by its employment so dramatic as they would appear intuitively. The information environment in America had already been radically altered in the decades preceding the diffusion of the telegraph. Compared with the acceleration in communications in those decades, the time-distance saving through use of the telegraph was small. For example, the average time lag between an event in Washington and an account of it being published in Boston was reduced by 11.8 days between 1790 and 1817 (from 18 to 6.2 days). Between 1817 and 1841 the lag was reduced by another 3.4 days, to 2.8 days.[3] In 1846, the telegraph virtually nullified, overnight, the time lag in communication of signals between Boston and Washington.[4] Although percentage-wise the telegraph reduced the time-distance between the cities infinitely, as a result of improvements in transportation since 1841, the absolute saving in time was only twenty-four hours.[5]

In the decades before the advent of the telegraph, what is commonly referred to as the "transportation revolution" was beginning to take place in America. The introduction of new means and processes of transportation enhanced the volume, lowered the cost, and greatly increased the speed of moving people and materials. By the time the telegraph arrived, long-standing notions of time and space were changing, and American society was adjusting to new standards of speed. The way in which the telegraph was first perceived, the patterns which developed for its employment, and the institutional role it eventually played can be properly understood only in the context of these developments.

"We stop the press to announce that there is no news; and none is expected. More anon!"[6] Similar, though usually more prosaic, announcements were common in American newspapers in the early nineteenth century. One enterprising editor, on an occasion when news was not forthcoming, printed President Washington's obituary from the previous century. The same editor occasionally inserted chapters from the Bible to fill voids resulting from lack of news; he justified his policy on the grounds that to nine-tenths of his readers these passages were news. The paper clearly anticipated the opinion of the yet-to-be-born *Christian Science Monitor,* whose motto proclaimed that "the news is news until it's read."[7]

The general predicament of turn-of-the-century journalism was the scarcity of news. The journalist Charles Congdon, skimming a volume of the newspaper he had delivered in his youth in a New England town early in the nineteenth century, found he was

> more surprised to find how little there was in it than to discover how little it was in itself. In that time of small things, subscribers must have been easily satisfied . . . The news from Europe, when there was any, was usually about six weeks old, or even older. Of editorial comment upon men and things, there was a plentiful lack. There were very few communications [that is, letters from correspondents] fortunately, for most of them were far from interesting. The advertising amounted to nothing at all.[8]

Later, he related that in those "earlier, simpler times . . . a good bit of news lasted a long time." In the circumstances, "an earthquake was a godsend" and other unique occurrences were "paragraphed and articled to death."[9]

The *Centinel of the North-Western Territory,* published in 1794, contained about one-seventh as much news matter as a regular issue of the *Cincinnati Gazette* of 1835. In 1816, the two most popular New York dailies contained on average between one-half and two columns of news.[10] The scarcity of news reflected a combination of the absolute dearth of events and occurrences, and the effects of prevailing conventions and standards of assigning news value to events. Human interaction—political, commercial, and social—which generates the bulk of newsworthy events, was far less than in subsequent decades.[11] There was less happening within a smaller population and over the narrower expanse of turn-of-the-century America than in the larger, more populous, and more urbanized nation of later years. Then too, communities in those times of relative isolation were more self-contained and inward-looking. In such a localized society, news of other areas and of common causes was less appealing than in a more centralized state, where the organic nature of national society made non-local information more relevant.[12]

Yet the fragmentation and localization of society did not generate an emphasis on local news. In fact, local news was seldom published in small towns, where local developments could be communicated by word of mouth.[13] Nor did "soft news" burden the pages of the newspapers of Congdon's boyhood. The human interest story—the accomplishments of amazing cats and dogs and such-like—was not considered newspaper material by the standards of turn-of-the-century journalism.

The island communities of the early nineteenth century did have a common interest in news of national and state affairs, most notably of debates and decisions in state capitals and in Washington. Indeed, newspapers of this period served mainly as sources of political information. They were commonly sponsored by parties that wanted their messages brought to the people. Yet politicians and parties were guided by a long-standing tradition that political debate and public business should be conducted intimately and orally, and that it "was not properly transacted in public."[14] Even after central party organs were established to supply readers in political centers and the country press with an official version of political developments, they obeyed the convention that it was best to publicize only a deliberate and very concise "skeleton of debate."[15]

The foreign news considered most noteworthy—that from Europe—arrived at intervals of several weeks at best, and it commonly dealt with events that had taken place several months previously.[16] Sporadic receipt of national news was just as characteristic, and it was caused to a large extent by the long intervals between mails. This infrequent movement of information directly affected the scarcity of news. Significant occurrences, even if short in duration, have a tendency to "develop."[17] They shed new light on the events and the conditions that preceded them and in turn produce consequences of their own. Thus, a fall in the stock market can snowball and end with a crash within several trading days. It may elicit bankruptcies, crimes, and administrative actions, while highlighting the political or economic developments that supposedly caused the decline. The more that time elapses between events and their report and the better developed the story becomes, the more the information is condensed and summarized. A report of the weekly change in the stock exchange contains all the information included in five or six separate daily bulletins. In the early nineteenth century, the long intervals between news reports thus made for economy of available news.

Besides the quantity and frequency of news, the velocity of its transmission was crucial in shaping news flow early in the century. If "old news is no news" or, as Robert Park put it, old news is "mere information," discussing the celerity of news in the period would constitute a misnomer. A great length of time was consumed in the process of communicating the news. In part, the fault for the slowness with which the public received news was that of lethargic editors. In a celebrated incident, an enterprising reporter suggested that the editor of the Boston *Mercantile Journal,* John S. Sleeper, send him to cover a speech Daniel Webster was to make in a neighboring town. Captain Sleeper declined the offer of his wide-awake reporter, saying, "Somebody will send us in something about it in two or three days."[18] Certain papers in rural New York were so lax in handling the news they received that "whenever the 'boss' or the boys had been lazy or frolicking too much" they resorted to their largest types in order to fill the requisite column or two of news, leaving the rest of the news to be written up for future issues.[19] The publication of the *Arkansas Gazette* of 1823 was set for Tuesdays, to allow the editor the leisure to prepare news received on Sunday.[20] Sometimes the day of a weekly's publication was not even fixed. In the case of one newspaper, where "the work in the office was always behind hand, and . . . the paper was never out in season," the subscribers simply "waited for the paper until it was printed."[21]

Editors, by delaying the publication of a news item for one or two days, contributed only a fraction to its longevity. A long series of delays intervened between the occurrence of an event in one part of the country and its publication in another. The initial delay resulted from the time necessary for a newspaper to receive a report of the event, either through a communication or in person. This delay could extend to several days even when the event took place in a densely populated area not far from a newspaper office.

The source of most of the news published in American newspapers was other American newspapers. This singular condition was of course a function of transmission rather than of news gathering. Although each news item still had to be culled from real life to be made public, it was made known throughout the country via the circulation of the newspaper first reporting it. Since colonial times, postal policy was to allow editors to exchange their publications post-free. By incoming mail they received news accounts that papers in other regions had first collected and reported.[22] Thus, to a large extent, it was the speed of the mails that regulated the pace of American news flow. Such speed, in turn, depended on the rapidity of post office handling at both ends, on the swiftness of the post office's mode of transportation, and on its frequency. The time consumed by the journey, and the wait for the next departure of the delivering transport, determined the interval between the occurrence of events and their being read about by newspaper subscribers.

The frequency of newspaper publication had an important secondary effect on the pace of news flow. A morbid anecdote, concerning a provincial editor in the West, illustrates this effect. The editor, on one bleak Monday, discovered the body of a neighbor who had committed suicide by hanging. He left the body under lock and key for two days, until after the publication of his competitor's weekly, on Wednesday, and reported it in his own weekly, on Thursday. "Do you think I was going to say anything about the suicide and let the scoundrel have the paragraph?" the editor supposedly asked.[23]

The infrequency of issue of newspapers, the source newspaper as well as the end-publisher, was a significant hindrance to rapid circulation of news. A majority of newspapers in the nineteenth century were published weekly or even less frequently, and the number of American towns (or for that matter, in the early nineteenth century, even states) in which dailies were published was minimal.[24] In the case of weeklies, a news item would make its first published appearance an average of three and a half days after it had been procured; and sometimes seven full days would elapse between receipt of information and its first publication.

The via dolorosa of the news item from its occurrence to its receipt by the reader was not over until the final editor used his scissors and paste and published the item in his next edition. Then, readers who lived some distance from the newspaper office had to wait for the mail to deliver the paper. In many cases, the flow of news from source to end-newspaper was more than a one-step process. Having traveled across the country from one bilateral exchange to another, the credit line of a news story could sometimes resemble the scoreboard of a long-distance relay race. The considerable time expended in that process of transmission and the leisurely pace at which each stage was executed make it quite difficult to associate the flow of news in that period, even remotely, with the latterday notion of a race for news.[25]

During the second quarter of the nineteenth century, America witnessed an exponential increase in human interaction within a growing population distributed over its expanding territory. The peace following the slump caused by the War of 1812 initiated an outburst of commercial activity. A dramatic expansion in trans-Atlantic trade energized the whole economy. The increase in business activity was facilitated by a revolution in transportation. Turnpikes were built, canals dug, and steam power harnessed—first to move vessels, then for the locomotion of wheeled vehicles.[26]

More frequent encounters between goods and markets, between letters of credit and merchants, and among producer, importer, and consumer increased the volume of information that was of interest to the business community. The newspaper now served as an important source for commercial information, in addition to its established function as a purveyor of political intelligence. At the same time, as political debate was becoming more open, the circulation of political news increased significantly.[27] The general increase in population and its greater mobility tightened social contact across space and enhanced the potential quantity and relevance of information from afar. An increasing population within urban centers created large concentrated markets for information.

The same developments in communications transformed the potential and the practice of news transmission. The increased frequency of departures and arrivals and the shortening of intervals between them resulted in a greater quantity of more timely information being transmitted more frequently from place to place. Consequently, the way the public experienced news was radically changed. "News" connotes both important events which have just recently occurred and the report of those events. The time difference between the two was growing smaller.[28]

Newspaper content in the second quarter of the nineteenth century reflected the reduction in delays in the various stages of the transmission process. In the period spanning 1820 and 1832, up to 28 percent of news items published in a sample of American papers reported events that had taken place a month or more previously, and only 21 percent of news items related to events that had occurred within the preceding three days. From 1833 to 1846, a corresponding sample indicated that only 13 percent of published news items concerned events more than a month old, and twice as many news items (41 percent) were of events that had taken place within the three days prior to publication.[29]

The sharp decrease in time lag between events and their report in out-of-town newspapers between 1790 and 1841 was just as striking. The most marked decrease was for the flow of information between the Midwest and New York between 1817 and 1841, as depicted in Table 1. This decrease reflected the three main sources of delay discussed above: length of time and frequency of transportation; time consumed in gathering the original news; and the delay owing to the process of publication. The only source of delay not reflected in these figures was the time necessary for delivering the newspaper from the printing press to its readers.

The effects of the transportation revolution on the first of these factors—the speed of transmission—may be best illustrated by contrasts in the speed of inland navigation between the early and mid-nineteenth century. In the latter period, fast steamboats plied the waters of the Hudson at 20 miles an hour. They could even complete the up-stream voyage from New York to Albany, a distance of 150 miles, in less than eight hours.[30] Several decades earlier, packet boats ferrying between Richmond and Lynchburgh on the James River had been liable to fines if they exceeded 4 miles per hour.[31] Early in the century passengers between Albany and New York spent from two days to two weeks on sailing packets; by the 1830's such journeys could be completed in ten hours. The introduction of steam navigation on the Mississippi reduced

Table 1 Decrease in Mean Public-Information Time Lag from New York to Other American Cities, 1794, 1817, and 1841 (in days)

City	1794	1817	1841	Decrease 1794–1817	Decrease 1817–1841
Charleston	22.5	8.2	5.5	63.6%	32.9%
Savannah	26.5	10.2	6.3	61.5%	38.2%
New Orleans	—	27.5	10.6	—	61.5%
Buffalo	13	5	—	61.5%	—
Detroit	—	18	7.5	—	58.3%
Cincinnati	44.5	19	7	57.3%	63.2%
Chicago	—	43	10.5	—	75.6%

Source: Allan R. Pred, *Urban Growth and the Circulation of Information: The United States System of Cities, 1790–1840* (Cambridge, Mass.: Harvard University Press, 1973), map 2.5, p. 41; map 2.6, p. 44; map 2.9, p. 51; table 2.4, p. 56.

the up-stream voyage from New Orleans to Louisville from three months to eight days or less.[32]

Land transportation was similarly speeded up, even before the spread of railroads, by advances in stagecoach technology and the construction and improvement of turnpikes. In 1800, travel between New York and Boston by stagecoach, the fastest means of transportation between those cities, took some seventy-four hours. By 1825, the journey required only forty-one hours; and in 1831 a traveler could reach Boston approximately thirty-three hours after leaving New York.[33]

Newspaper proprietors benefited from these developments, which helped hasten the national flow of news. Publishers, however, also contributed to the process, by increasing the frequency of their publications' issuance. The publisher had an obvious interest in increasing the frequency of his periodical. Assuming that he could retain most of his subscribers while multiplying the number of issues, his enterprise would expand many-fold. The idea must have crossed the minds of several editors prior to 1782—when the first American daily reportedly appeared—and many other minds subsequently. Most newspaper publishers probably would have succumbed to the temptation to publish daily, but for the lack of sufficient raw material in the form of news.

With the general acceleration of the pace of life, as well as the increased net volume of information and its more frequent transmission, the editor had sufficient subject matter to make a daily edition possible. The developments in transportation, at the same time, enabled him to dispense his now daily perishable product, while still fresh, to a larger audience at an expanding radius from his printing shop.[34]

The number of daily newspapers published in the United States increased in proportion to the improvement in communication facilities. Between 1800 and 1810, while the population grew by 36 percent, only two dailies were added to the twenty-four already available. During the subsequent decade, however, the number of dailies increased by 62 percent. The 1820's witnessed an increase of another 55 percent; and in the 1830's the number of dailies more than doubled (112 percent), with a further increase of 84 percent in the 1840's. Although still rapid, the rate of growth decreased in the 1850's and 1860's, to 52 percent and 48 percent, respectively. The increase in the proportion between the number of dailies and the population followed a similar pattern.[35]

Newspaper editors at the time also contributed to accelerating the flow of news in a more direct way. They expended great efforts to collect news rapidly and invested heavily in speeding its transmission and publication. A significant movement to speed the news emerged in New York City in the late 1820's, led by the newly established *Journal of*

Commerce and the recently consolidated *Courier and Enquirer.* The two newspapers, trying to attract the patronage of the commercial community, made a specialty of speeding business news and breaking political stories to their readers. They competed for swift transmission of Washington news by establishing rival pony expresses between New York and Philadelphia—later even Washington. They also competed for early receipt of foreign news and shipping reports by employing fast boats to intercept vessels entering the city's port. Procuring timely shipping news cost each of these papers between $15,000 and $20,000 a year. Between 1828 and 1830, the *Journal of Commerce* expended $30,000 on its news expresses alone. The paper had been purchased outright for a lesser sum just prior to these arrangements.[36]

Another surge in efforts to get the news rapidly began in the late 1830's, after a new group of dailies in the major northeastern cities—the penny press—became securely established and prosperous. They entered the competition for early news with much energy and zeal and with enormous financial resources. Uncounted thousands were expended in this outburst of news enterprise. Outlays on the order of $1,000 were not unusual for expressing single news items. The firm of Abell, Swain, and Simmons paid $2,000 for a fleet of carrier pigeons to speed news from Washington to the Baltimore and Philadelphia penny papers they owned. On one notable occasion, several enterprising newspapers even joined together to charter a special schooner for a trans-Atlantic voyage to hasten the receipt of European news. A number of New York dailies that combined in the 1840's to express steamer news from Halifax to New York spent $1,800 for each crossing of the Bay of Fundy by chartered steamer.[37]

The *enfant terrible* of metropolitan journalism, James Gordon Bennett, founder of the *New York Herald,* provided the prime illustration that time was money in the press business. He contracted to pay Daniel H. Craig, an innovative news entrepreneur, a $500 bonus for each hour European news sent by Craig to Bennett's *Herald* preceded receipt of the same news by the *Herald*'s New York competitors.[38]

Editors of what came to be known as the "penny press" were the leaders in expediting the transmission of news. This group of editors in fact claimed credit for introducing "energy" and "enterprise" to American news gathering. Their contemporary commentators generally accepted the notion that the penny press led the way to discovering the merits of timely news and agreed that in the process they revolutionized journalism.

James Parton probably overstated his case in calling publication of the first penny newspaper the event that had the "greatest and most lasting influence" on the American Republic in its first seventy years.[39] Even if not the "one event which stands out so prominently beyond and above all others," the emergence of the penny press in the mid-1830's represented an important, perhaps even the most significant, departure in the history of American journalism. The early penny press promoted the speed of news at any cost.[40]

As a group, the proprietors of the first penny papers did not belong to the social elite of New York. Most were practical printers, a vocation that had no claim to social prominence.[41] Characteristically, they were not well connected, either economically or politically, and their social distance from sources of economic and political power determined, to an important extent, the fields they covered and the way they covered them. The original proprietors of the penny newspapers could not rely on communications from friends holding key economic positions, since they had no such friends. Nor could they depend on information supplied by the party establishment to political insiders, which they clearly were not. These aspiring editors had to gather social, economic, and political information from the outside, by independent efforts. In practical terms this meant they had to hustle around town to find filler for their news columns, and the easiest information to come across was local news, which was also the "fastest" news. Reports of crimes and sermons, of fires or parties, of trials and public meetings, of a sudden death or an engagement notice disclosing a discreet courtship could be collected instantly by merely walking around town. Yet these items interested readers. In gathering local news enterprising editors expanded the field of human activity that they scrutinized in search of recent developments. While this helped make the newspaper news-rich, almost by default it made for added timeliness of newspaper news.

Even the political independence characteristic of the penny papers carried a bonus in terms of news value and timeliness. Although most of the penny papers were partisan, their relationship to party was different from that of their predecessors. Because they were commercial institutions, established, financed, managed, and edited through private and not party enterprise, the loyalty of even the most party-committed of the penny papers had its limits. Their readers could therefore expect now and again a refreshing dissent from the strict party line.[42]

Editorial opinion applying party doctrine to current events had previously developed out of consultation with political leaders, and its phrasing frequently required approval by party officials. This was time-consuming and caused a lag between news and editorial comment.[43] For example, Jefferson's first annual message to Congress, delivered on 8 December 1801, was published by the *New York Evening Post,* at that time a Federalist organ, on 12 December. Not until the 17th did the newspaper print its comments on this first official utterance by the great political rival. The New York Federalist contingent, headed by Alexander Hamilton, obviously needed the interval to devise an authoritative statement of opposition policy. By the 1830's, a committed editor of a penny newspaper, not directly dependent on the party, could have his editorial say on the most recent developments within hours. Thus, the editorial section, a large part of the newspaper, adopted the more rapid pace of news copy, and the newspaper as a whole seemed more up-to-date.

The human interest story, serial fiction, hoaxes, and similar features, considered important innovations of the penny press, probably emerged as an answer to the needs of independent editors for fresh copy. Such non-news items had the advantage of creating suspense, similar to the effect of developing news. Human interest stories, which had some of the attributes of real news, may not have represented an achievement in timeliness, because they could have been written weeks before they were printed. Yet they had the merit of not being dated and could be made to appear as recent occurrences. However significant these journalistic innovations were, they were less important in promoting the timeliness of news than the revolutionary marketing system the penny press adopted.

Traditionally, a considerable capital reserve and a lengthy list of pledges to subscribe were considered prerequisites for launching a new newspaper. Those who could afford to subscribe to a paper, let alone invest in one, belonged to the more affluent classes. The early commercial press was supported by the business community, and the political press depended on party support. Political influence was often instrumental in raising funds from the party faithful and in convincing those who could afford it to subscribe. Political leverage was important also for sustaining the newspaper once it was established; parties mustered support in legislative and administrative bodies for subsidies to loyal newspapers, usually in the form of inflated contracts for public printing.

The early penny press proprietors could not expect much financial aid from commercial circles, or support from influential politicians. With precious few reserves of capital, their newspapers had to become viable commercial enterprises if they were to survive the first few weeks of publication. These obscure printers and journalists had to devise novel strategies for making their ventures instant commercial successes. The most crucial of these was a new marketing strategy.

According to legend, the idea of a penny newspaper originated with a medical student, Horatio Davis Sheppard. Walking down busy Chatham Street in the New York of 1830, Sheppard, impressed by the variety of the products vended for a "penny a piece," noted how readily people parted with the smallest currency denomination. Then and there he came up with an idea: why not sell newspapers in the streets, at a "penny a piece"? The remainder of Sheppard's daily stroll was devoted to crystallizing the idea of a daily newspaper that would share the marketing miracle of pieces of cake, single apples, or pairs of shoelaces.[44]

Finding an alternative to the traditional channels of newspaper distribution was indeed a revolutionary undertaking. Prior to 1833, when the *New York Sun*, the first successful penny newspaper, was established, consumers encountered newspapers only on their doormat. They were purchased by means of a long-term subscription and were received by the subscriber at his home or office. "Newsboys are unknown in 1833," wrote an employee of James Gordon Bennett on the penny *New York Herald:*

> At the steamboats, the newspaper carriers are finding that they can sell a few copies, at six cents each—probably to hungry politicians, who are hurrying away to Albany or to Washington; or, perhaps, to merchants who wish to see the ship news, or the prices current . . . The publishers will print, for the carriers, only a few extra copies. It is quite uncertain if the papers can be sold, even if they contain libels on the President, or his family, or on a score of politicians. A first rate murder, well described, may make the sale of a hundred copies quite possible![45]

The penny papers adopted the London plan of newspaper marketing. Distribution was delegated to retailers, who purchased the newspapers on a cash basis. These retailers could either sell them to subscribers or hawk them in the public domain. In the former case, given the cash basis of the arrangement, subscriptions were usually for short periods of time, characteristically a week. The street sales proved very successful, and newspapers recruited teams of newsboys to hawk their product.[46]

This new system of distribution appears to have had a decisive impact on the time frame of news. Experience, if not common sense, dictated that the news content of the individual edition would determine its sale in the streets. As seen by Isaac Prey, an active journalist of the period, the predicament of the penny press was that "the public is a fickle monster—may buy, and may not." Because by necessity "the people must be kept wide awake with something," it was only logical to conclude that "news must be forthcoming at all hazards, or why have a newspaper!"[47]

To attract the occasional reader, and to keep the short-term subscriber content and loyal, editors made sustained efforts to keep readers rapidly informed. A dramatic scoop—or "news beat," in contemporary terminology—could even convince a subscriber to one of the established "blanket sheets" or, for that matter, to a competing penny paper to spend one more cent and buy, on the street corner, an additional newspaper containing the latest news. After all, the spontaneous, inspired passerby had a streak of what a Boston editor of the 1810's considered "one of the peculiar traits of national character." This was "the insatiable appetite which exists in all classes of people in this country for news. It is a thirst so universal that it has given rise to a general and habitual form of salutation on the meeting of friends and strangers, *What's the news?* This is an inquiry of such universal interest that he who can answer it is always welcome, while he who brings the second report of an event, although it be much more full and correct in its details, is listened to with indifference." The penny editor, in order to be "always welcome," had to "keep pace with the curiosity and anticipations of a great part of his readers." No sophisticated machinery of market research was necessary for this understanding. As Horace Greeley put it succinctly, "The public are apt to take the paper which has the first news."[48]

The innovations in distribution facilitated the emergence of the "extra" as a regular institution. Once the penny papers achieved a significant following, they had in place a network of newsboys available for street sales throughout the day, and their printing establishments were structured to produce a massive number of sheets in a minimum amount of time.[49] Coming out with an extra was prestigious, because it implied a news beat; and it was remunerative, because it sold additional newspapers—many thousands in the case of important news.[50] Besides, if exclusive, an extra could frustrate and damage competitors, since it made the regular morning editions still on sale on street corners as obsolete as "yesterday's news." The possibility of publishing an extra provided editors with an added incentive to speed the news; now every hour could produce a news coup, rather than the full twenty-four hours between editions, as had previously been the case.[51]

By the mid-1840's extras became so prevalent that they were virtually institutionalized as semi-regular late editions.[52] As Henry J. Raymond once said, tongue in cheek, "The press had ever been reluctant to put itself before the public; no paper ever appeared more than three times a day."[53] New Yorkers were bombarded more and more frequently with breaking news stories in extra editions, with the result that the public received ever more timely information.

Newspaper editors used these extras for self-promotion. Moses Y. Beach of the *New York Sun* continually delighted himself and perhaps his devoted readers, too, with minute descriptions of the beats his paper achieved over its established competitors. "Who will wonder, after this," glowed the *Sun* after an important scoop, "that the lazy lumbering lazaroni of Wall Street [the six-penny newspapers] stick up their noses at us?" The paper relished describing its news arrangements, and it introduced one account by asking: "Did our contemporaries believe that we ever could be at fault in furnishing the earliest news to our readers? Or did they indulge the hope that in newspaper enterprise they could ever catch us napping? . . . Our extras are always out first."[54]

The race for the news was in itself a good story. It had the exhilarating attribute of a clear, uncontested result with winners, losers, and rank order finishes for those in between.[55] Strenuous competition for supremacy in news was novel, and it apparently appealed to the public. After the penny papers became affluent establishments, they could afford to adopt the most rapid means of transmission that transportation technology made possible. These novelties considerably increased the public's fascination with the competition for timely reporting.

The means of achieving the news beat added to its importance in promoting the newspaper. As early as the 1820's, a semaphoric telegraph operated by or on behalf of the *Journal of Commerce,* between Sandy Hook and New York City, informed the *Journal* of incoming ships.[56] The telegraph's signals were just as visible to the general public, and when it went into action, it was "notifying the whole metropolis of the approaching news-boat." The effect of such an imaginative effort by the *Journal of Commerce,* claimed a contemporary journalist, was that "that paper not only by this means obtained its news ahead, but its enterprise advertised it all over the city and country."[57] By the time the penny press moved into the vanguard of news gathering, it was the pony express, steamboats, and locomotives that provided the excitement.

The public in the first half of the nineteenth century was fascinated and provoked by speed and the new devices that were introduced into the struggle for timely news.[58] One eyewitness to a competition between pony expresses in the 1840's recollected that "there was almost as much excitement in the village over a race between the two expresses as there would be in a general election at the present time [c. 1912]." The observer distinctly recalled "numbers of people standing in the street, waiting to catch a glimpse of the riders as they approached."[59] The winners of the competition for news received great publicity. The whole process and atmosphere of the competition, however, contributed to public awareness of, and interest in, news; hence, all participants in the race for news were winners.

The universal usage of the terms "penny press" or "the cheap press" underscored the importance of the revolutionary price policy that the new dailies of the 1830's introduced as part of their marketing strategy.[60] It was of course the low price of the newspaper that helped translate whatever attraction timely news held for the public into large newspaper sales. The cheap goods strategy in marketing the daily to consumers was, however, augmented by a premium strategy in pricing advertisements. Advertisers who realized the great power of the cheap press to promote products and services among the rapidly growing population of urban centers ultimately footed the bills for operating expenses as well as allowing enormous profits for penny press proprietors. From the advertisers' point of view, the penny papers were anything but "the cheap press."

Soon after the circulation of successful penny papers reached unprecedented heights, advertisements began pouring in, and gradually, the price of ads was adjusted to the number of copies that the newspapers sold. By the late 1840's, a firm could pay $60 a year for advertising in the prestigious *Journal of Commerce,* but $1,000 for a similar display in the *Herald.*[61] Because advertising revenues were directly related to circulation, and proprietors recognized that timely news attracted readers, revenues from sales and from advertisements were reinvested in the expensive machinery necessary to increase the celerity of news. As Harold Innis observed, news became "a device for advertising the paper as an advertising medium."[62]

With mass circulations justifying expensive ads, the customary long-term advertising contracts gave way to short-term promotional campaigns, often to one-time displays. Consequently, the advertising pages became more varied and dynamic, contributing to the overall freshness of the newspaper. By 1848 the *New York Herald* required its advertisers to change their ads daily. It explained to its patrons that "advertisements in the *Herald* are read as much as items of news; but the lumbering and repeated notices in the large papers [blanket sheets] are looked at and never heeded after the first day."[63] Even Horace Greeley regarded advertisements as a "portion of the news of the day."[64] Thus, the advertising pages were assimilated into the dynamic tempo of the penny paper's news columns and contributed to the overall timeliness of its copy.

Circulation figures for daily newspapers in the first half of the nineteenth century are sparse. The fragmentary data that exist demonstrate that the penny press achieved a tremendous following within a remarkably short period of time. The circulation of New York dailies more than tripled between 1832 and 1836 (from 18,200 sheets to 60,000).[65] In that four-year period the *New York Morning Sun* made its first appearance, and in rapid sequence the *Transcript,* the *Man,* and finally, in 1835, the *Herald* (among numerous other short-lived dailies) were competing with it at one cent a copy. Most of the added patronage of dailies in the mid-1830's went to these papers. By the end of 1835 the circulation of penny papers in New York (28,500) was more than double the circulation of the six-penny press (12,500).[66] The two-to-one ratio between the circulations of the two types of dailies persisted in 1842 even after some of the cheap papers raised their price to two cents and the overall press circulation had approximately doubled, as illustrated in Table 2.

The great increases in circulation must have reflected to some extent the attractiveness of timely news for the reading public.[67] Circulation figures of New York dailies demonstrated an additional marked expansion in the late 1840's, when the telegraph was applied to the rapid transmission of news.[68] The timeliness of reported news was closely related to the increase in public interest. Through the accelerated pace of news gathering and transmission, newspaper intelligence was being transformed; from a history of tidings of the past it was becoming a chronicle of the present.

Table 2 Circulation of New York City Dailies in November 1842

Newspaper	Circulation
Penny papers	
Sun (1¢)	20,000
Herald (2¢)	15,000
Tribune (1.5¢)	9,500
Aurora (2¢)	5,000
Chronicle (1¢)	5,000
Morning Post (2¢)	3,000
Plebeian (2¢)	2,000
Tattler (1¢)	2,000
Union (2¢)	1,000
Total	62,500
Six-penny papers	
Journal of Commerce	7,500
Courier and Enquirer	7,000
Express	6,000
Commercial Advertiser	5,000
Evening Post	2,500
American	1,800
Standard	400
Total	30,200

Source: Frederic Hudson, *Journalism in the United States from 1690–1872* (New York: Harper and Brothers, 1873), p. 525.

By accelerating the speed of news, editors revolutionized the operational utility of newspapers to their readers and thus attracted more of them. The ability to monopolize news is a function of time; as James Carey put it, when information "moves at unequal rates of speed, what is already the past for the privileged is still the future for the deprived."[69] When newspaper news traveled at the speed of the mails, information gained from private correspondence was usually more timely. Information contained in a letter could be acted upon at once, while that destined for the newspaper was being edited, printed, and distributed.

Traditionally, therefore, newspapers did not even try to compete in timeliness with the person-to-person flow of information. Boston papers, considered the most enterprising in the collection of marine news in the 1810's and 1820's, received much of the data they published from "late letters just received by merchants."[70] Similarly, commercial quotations published in Boston's leading *Columbian Centinel* were "corrected by two brokers weekly," prior to publication.[71] Under these conditions, information advantages were enjoyed exclusively by a limited circle of operators. A well-connected merchant, through his established business ties, could receive by mail or by envoy key mercantile information ahead of his less-established or less well connected competitor, and certainly ahead of the general public, who depended on the newspaper.

Private monopolization of timely information disappeared with the emergence of news enterprise by journalists.[72] The journals competing for news supremacy in New York directed most of their efforts to expediting the receipt of news from Europe and Washington. Prices in American markets fluctuated according to changes in foreign, most notably English, markets, and they were influenced to a large extent by economic and political conditions there. The early receipt of European news therefore implied enormous economic power. Washington news—and, to a lesser extent, Albany news—carried great political significance and was important from a commercial standpoint as well.

Initially newspapers employed private expresses to hasten the transmission of news from Washington—first the pony express, later the locomotive express. In the former case, swift, full-size horses were dispatched in relay. In the latter, locomotives were chartered for special runs. Horses were most effective for short distances, with fresh riders, and of course only when unencumbered by heavy loads, let alone a coach. The rail express required a heated locomotive ready to run on an open track, with synchronized connections on other rail lines at the ready.

The enormous expense of running an express, and the arduous task of organizing one, made it available only to enterprises with large budgets for speeding news, such as each of the major New York dailies. The great effort and high price necessary for an express system was most cost-effective if used repeatedly. It was barely productive, if at all possible, for one-time efforts. The press was the one sector that had a constant interest in, and sufficient funds for, sustaining a continuous flow of speedy news. It was therefore the press that pioneered in employing these advanced devices for speeding the pace of news. As the *New York Herald* claimed, had the press not resorted to expressing news, "the public would be to this day trusting to Uncle Sam's mail-bag for the earliest intelligence."[73]

Even more complex facilities were necessary for speeding news from overseas markets. Unlike congressional and political agendas, which arrived with regularity, trans-Atlantic transports carrying the latest European news were unscheduled. Even with the development of semi-regular packet lines, the expected time of arrival could be established only within a matter of days. There were also enough unscheduled voyages, with each arrival possibly bringing crucial economic and political data, to warrant constant watch of the harbor offings for the latest arrivals. The time-consuming process of docking and later of quarantine suggested to enterprising news-gatherers that they could save time by pre-boarding vessels and stripping them of their news cargo prior to their entry into the harbor.

By the first decade of the nineteenth century, harbor information brokers had established themselves in the major American ports. Early ship news was brought into offices established for this purpose, where it was entered into books open to the scrutiny of subscribers. The names of these establishments, such as Merchants' News Room, or Exchange Coffee House, clearly indicate who the patrons of these facilities were, and who profited from the monopoly in time over European news.[74]

It was not long, however, before the newspaper became the recognized purveyor of the latest commercial information. An eyewitness described "crowds, in Topliff's News-room in Boston, disagreeably elbowing each other around the file of the *Journal of Commerce,* on the arrival of the New York mail, for the brief one and two line items in the late editions." He also recalled having seen David Hale, the proprietor of the *Journal of Commerce,* "mount a chair in his office in New York and read off to the crowds of merchants, with the voice of Stentor, some fresh news just received." Moreover, when the semaphoric telegraph had announced that the *Journal of Commerce*'s news schooner was approaching the harbor with the latest news from Europe, "business would be at once suspended to await her arrival."[75] By the 1840's it was thousands of New Yorkers who found the timely acquisition of news from distant markets worth hours of waiting in front of newspaper offices. On one such occasion the *Herald* commended the "patience" of the "thousands who had congregated about our office" before their "intense anxiety" was relieved by the appearance of an extra.[76]

As newspapers gained control of the fastest possible means of transmitting foreign news and Washington news, they achieved a virtual monopoly of this sensitive knowledge.[77] With great pride they now presented their most recent news items with headings indicating how many hours or even days ahead of the mails the information was. Early news procured through the elaborate arrangements of the New York press was regularly posted on bulletin boards outside newspaper offices.[78] It is questionable whether this practice encouraged the immediate sale of newspapers. Although some must have been encouraged to purchase a paper to receive the full report of the information, others were probably content with just jotting down the data important to them. Be that as it may, the practice surely was important in advertising the newspaper as the source of the earliest news.

Publishing the latest news on bulletin boards was also symbolically significant: it highlighted the fact that the most sensitive information was now in the public domain, open to scrutiny by all. "Its importance cannot be estimated," James Parton suggested of the penny press, as "it brings the Course of Events to bear on the progress of every individual. It is the great leveler, elevator and democratizer." James Gordon Bennett summarized the same achievement of the penny press in breaking a monopoly of knowledge rather tersely: "Speculators should not have the advantage of earlier news than the public at large."[79]

By 1844, the great achievements of the press and the public in the struggle for control over the latest news appeared to be seriously endangered. The telegraph made its premiere, and the success of its preliminary performance indicated that before long it would become a force to reckon with in news gathering. The potential of the telegraph to provide speedy news simultaneously throughout the land surpassed all previous means of achieving timeliness in news. Editors came to realize that the news business would never be the same. On 7 May 1846, after excited competition between expresses for the publication of the foreign news, the *Herald* concluded the description of its own beat on a somewhat solemn note. "[Our] express from Boston, which brought the first intelligence of the arrival and going ashore of the *Cambria,* will probably be the last express of the kind which will ever run between the two cities, with foreign news. *Vale, vale, longum vale!*"

The *Herald* was bidding farewell to the dearly achieved monopoly of the press in supplying the most timely information.

2

"Time Past to Time Present": Linking the Telegraph and the Press

Samuel Finley Breese Morse didn't attend his wife's funeral.[1] Early in February 1825 she fell ill in New Haven, following complications during the birth of her fourth child. Morse was in Washington at the time, on a lucrative commission to paint the portrait of the Marquis de Lafayette. It took a number of days for the news of his wife's deteriorating health to reach Morse in Washington, and while preparing to return home, he received a letter informing him of her death. It reached him five days after the fact. When he finally arrived in New Haven, all that was left to do was to comfort four miserable children the best he could.[2]

This tragic personal experience, and the commuter marriage that preceded it, may have been enough to motivate Morse to chip away at the time necessary for communications to travel through space. In addition, his background made him uniquely conscious of the dimensions of time and space. His father, the Reverend Jedidiah Morse, was to the teaching of geography in America what Noah Webster had been to American spelling. His labors in researching and writing the standard school textbook of geography must have given his son great exposure to the concepts of distance, direction, shape, and area.[3] Landscape painting, in which Morse excelled, sharpened his perceptions of spatial relationships.[4] In 1832, on board a vessel trying to make the best possible time between England and the United States, Morse began considering ways of eliminating time in communications between individuals across space. The virtual simultaneity of the electric current gave him his solution. But he was not alone. No fewer than sixty-two competitors came up with ideas for the practical application of electricity to communication, comparable to those of Morse, at about the same time.[5]

Speed was on the minds of Americans in the 1830's and 1840's, and a revolution in the perceptions of time and space was in the making. Old notions of the limitations of speed were disappearing. Americans experienced in rapid succession a number of revolutionary innovations in speeding movement through space, among them the steamship and the railroad locomotive. Their ultimate commercial success made further discoveries appear promising endeavors to aspiring inventors. Gradual developments in the experimental study of electricity led many to turn in that direction for comparable devices.

Inventing the telegraph was perhaps less difficult than popularizing it. There were three important stages in developing the idea of the telegraph into a widely used and successful device. The first was simply to convince people that an electric telegraph could work and to demonstrate that it actually did work. The second challenge was to prove that the telegraph, besides being a delightful contraption, was also a useful instrument, with the potential of becoming a great means of communication. The third was to find enough users to make it a viable enterprise and a paying proposition for investors.[6]

Partly by chance and partly by design, in all three of these challenges the transmission of news played an important role. News demonstrated the capabilities of the telegraph and proved its usefulness. Moreover, the press became the telegraph's first great patron and figured prominently in bringing about its widespread diffusion. News transmission thus performed the role of the "supervening necessity" in the coming of the telegraph.[7]

Before the press took an interest in telegraphic technology, Morse experienced immense difficulty in diffusing the invention. Only after years of lecturing, lobbying, and negotiating was a bill to appropriate funds for an experimental telegraph line brought before Congress, in 1843. In the course of debate, the bill was encumbered by a proposed rider, appropriating funds for experiments in mesmerism. This was perhaps less surprising than the fact that the legislators were discriminating enough to vote down the rider and uphold the original bill.[8]

To the uninitiated, electricity was no more than a "mysterious agency." Morse found it difficult to persuade even the best informed to support the demonstration of a process they did not fully understand.[9] Besides, the great previous developments in achieving speed had an effective physical representation. The steamship and the locomotive were real objects that produced odors, made noise, and, above all, moved at a speed that could be appreciably sensed. In contrast, the telegraph moved information over great distances almost instantaneously, without any visible movement.

Morse faced the additional hurdle of demonstrating that his invention worked. His challenge was to prove to those present that the clicking machines they were watching were actually responding to operations taking place miles away. It was imperative that Morse find a way to prove to skeptics not only that the telegraph could work but also that it actually did work. The transmission of news turned out to be crucial in solving this problem.

Morse was apparently oblivious to the depth of the skepticism surrounding his invention. He considered the appropriation of $30,000 for an experimental telegraph line full vindication of its merits. After the appropriation was approved, he gleefully demonstrated to a congressional committee the operation of his machine. The congressmen were instructed to choose a message, and the professor commenced sending it over a five-mile circuit. Immediately the receiver began clicking, and the stylus punched the chosen message, "Mr. Brown of Indiana is here," on a slip of paper. Triumphantly, Morse held up the slip to the congressmen in attendance and read the message. They were not impressed, not even Mr. Brown of Indiana. They saw only a line of neatly impressed dots and dashes. "It won't do. That doesn't prove anything," whispered one congressman. "That's what I call pretty thin," remarked another.[10]

At this juncture, Morse was aided by the appointment of an authorized skeptic. In the spring of 1844, the second assistant postmaster general, whom Congress had made responsible for overseeing Morse's appropriation, asked John W. Kirk to keep an eye on the "impracticable or crazy" painter. Kirk was to find out if the $30,000 Congress appropriated for "Morse's foolishness" was being properly spent.[11]

Initially skeptical, Kirk began to sympathize with Morse, and perhaps even became convinced of the practicality of his invention. He proposed a way to verify that the dots and dashes actually represented information transmitted from afar. The Whigs were shortly to convene in national convention in Baltimore, and when they nominated their presidential candidates, the ticket could be telegraphed to Washington. Because of the importance of the news, curiosity would lead even skeptics to take notice of the telegraphic version of the results. Best of all, by transmitting news of a public event, the truth of the communication would soon be verified.

On 1 May 1844, Morse's competent assistant, Alfred Vail, from a makeshift platform near Annapolis Junction telegraphed the results of the nominating convention to Morse, who was twenty-two miles away in Washington. A great crowd was assembled in the Washington depot of the Baltimore and Ohio railroad, eagerly awaiting the arrival of the train from Baltimore, which was bringing the news of Henry Clay's expected nomination. Fortunately for Morse, Kirk, and the second assistant postmaster general, the Whig convention chose a relatively obscure New Jersey politician as Clay's running mate. The nominations, as received by telegraph, were announced to the waiting crowd.[12] "It is easy enough for you to guess that Clay is at the head of the ticket," reasoned one spectator, "but Frelinghuysen—who the devil is [Theodore] Frelinghuysen?" An hour and four minutes later, the passengers who arrived aboard the Baltimore train confirmed the accuracy of the first news message transmitted by telegraph.[13]

Mainly because of the effectiveness of this publicity stunt, there was a considerable crowd on hand when Morse officially inaugurated the completed line extending from Washington to Baltimore, on 25 May 1844.[14] A query from the Book of Numbers, "What hath God wrought?," was chosen to inaugurate the American telegraph service. Baltimore responded to the eternal query instantaneously, with an anticlimactic "yes." But Morse had obviously learned a lesson from his earlier demonstration of the operation of the telegraph line. The second telegraphic inquiry of record, from Morse in Washington to Alfred Vail in Baltimore, read "Have you any news?"[15]

At the time, Baltimore was hosting the national nomination convention of the Democratic party. Once more, news transmission via telegraph could serve as proof of both the operation of the invention and its utility. Frequently, when large groups appeared in Morse's telegraph room in the north wing of the Capitol, he called telegraphically on Vail to send him news from the convention. Vail dutifully dispatched Ezra Cornell, who thus became the first telegraphic news reporter of record, to the hall of the convention to bring word of the latest proceedings to those waiting on Morse, forty-two miles away.[16]

Again, the climax came when it was time to announce the nominations made by the convention, on 29 May 1844. Morse, reading from the punctured slip, informed the excited throng of politicians surrounding him that the nominees were James K. Polk and Silas Wright. The results, this time, were accepted with surprise, but without much doubt. In the telegraph room or soon called in, Silas Wright, who was perhaps unhappy with the convention's nomination for president, was utterly opposed to its selection for vice president. He wished to announce to the convention his declination of the "second highest but least important" office in the United States. Morse obliged the veteran senator by wiring his message to the convention, which was still in session in Baltimore. There, too, the dispatch was deemed credible enough to appoint a committee to negotiate telegraphically with Wright, and through Alfred Vail they asked him to reconsider. He did not change his mind, notifying them telegraphically of his refusal.[17]

For Morse, this sparkling demonstration of the telegraph's potential probably represented two victories in one: it confirmed the utility of his invention for news transmission; and, no less important, it demonstrated its multidirectional capacities. Unlike most previous technological advances applied to communications, the telegraph was not a vehicle but a channel. The demonstration from the Democratic convention proved that persons at great distances from each other could interact in virtually real time in a format resembling a conversation.

Morse originally saw the telegraph as encouraging social intercourse through space. He had a vision of his invention's linking wives with their distant husbands, allowing children to communicate with their parents, and encouraging lovers to exchange sentiments over the wires. Thus, Morse considered the events of the Baltimore convention to have happened "most opportunely for the display of the powers of the telegraph," because the device was used as a "means of correspondence, in one instant."[18]

Both these triumphs were not lost on the press. Because of the obvious news value of this exchange to all Americans, the message publicized throughout the land the medium that transmitted it; the telegraph was soon dubbed "the great highway of thought," its wires "slender bridges." The story of the nomination and that of the telegraph became integrally linked, and newspapers across the country reported from Baltimore two dramatic news items in one.[19]

"That intelligence can be instantaneously transmitted to great distances by Morse's Electro-Magnetic Telegraph is no longer a matter of speculation, but of certainty," announced one editor. "The Telegraph between Washington and Baltimore has satisfied the most doubting." The demonstration, besides proving the telegraph was practicable, suggested its usefulness. "That this Telegraph can be applied to the practical business of life is now an established fact," announced the same editor.[20] More was necessary, however, for the successful dissemination of the invention. Predictions aside, it was still imperative to find paying users.

The obvious potential users included the government, business organizations of great scope (a category which in 1844 included very little besides the fledgling railroad industry), the commercial community, and the press. The first two of these sectors could use the telegraph to coordinate operations over space; the last two for the accumulation of knowledge. There was also the potential of extensive use of the telegraph by the public at large, for exchanging messages of a personal nature.

For business operations, and to a lesser extent government administration, the interactive capabilities of the telegraph were potentially useful. Efficient management and control of large enterprises implied a continual input of information from a large field to a center, as well as response by instructions from headquarters to the periphery. Two-way communication was practically essential for many kinds of social exchanges, yet in the early development of the telegraph, the device's potential for instant two-way communications was seldom realized.

During the first year after the government telegraph line was opened to the public, it was rarely employed for social exchanges. Of 366 messages transmitted between 16 April and 4 May 1845, only 21, or less than 6 percent, were either addressed to or sent by a woman. Assuming that in the number of messages of a personal nature there was not likely to have been a marked difference between the sexes, this finding may suggest that the telegraph was used only marginally for social intercourse.[21]

As for government, the postal service employed the wire only twice during that period, and the collector of Baltimore, once. On 20 March 1845, a Corporal Wood sent a message to a Captain Walker in Washington. Whether this message was personal or official is not known, therefore we cannot count it among the first applications of the telegraph to military business. Even the president of the United States received a telegraphic communication. It came, however, from a newspaper editor rather than from a member of his administration. If government used the telegraph sparingly, the Baltimore and Ohio Railroad apparently did not use it at all. No telegrams to or from officials of the railroad company appear to exist.

The major users of the telegraph line were members of the commercial community and of the press corps. Eighty-five messages, almost a quarter of the total transmitted during the period studied, were directed to or sent by commercial firms. In many other communications the names Tappan, Taylor, Cooke, Corcoran, and those of other leaders of the business community frequently appear. Merchants and commercial companies used the telegraph regularly. The Baltimore firm W. B. Pyfer and Co., for example, received at least one dispatch from Washington every day but one in the period studied. Their telegraph expenses amounted to $10.20, or 13.5 percent of the entire income of the line.

The other frequent users were the newspapers of Baltimore and Washington. The telegraph transmitted fifty-one messages for the press during the period, constituting 14 percent of the total number of messages. Press dispatches, however, were significantly longer than personal or business messages (press messages averaged forty-six cents, other messages twenty-four cents), and they accounted for 32 percent of the line's income. If we take into account that the price of transmission per word decreased markedly with increase in the length of the message, the press actually occupied the line much more than its share of the receipts—almost a third—would indicate.

The striking fact that emerges about the first telegraph line from the account books of 1845 was its under-employment: it operated at only about 15 percent of its capacity, and its wire was idle throughout most of the day. The business results of the first line clearly were not a good indication of the commercial potential of the telegraph. The Washington-Baltimore line did not connect the major urban centers of the country. Moreover, there was a certain incompatibility between the two cities linked by the first American telegraph. Washington was a political and administrative center, but not much of a city; Baltimore was a city, but not much of a center. Then, too, the distance between them was only forty-two miles, and a railroad was already operating regularly between the two. The longer the distance over which messages were transmitted, the greater the savings in time. With the development of the automatic repeater, a message from Washington to as far as Detroit could be sent as quickly as one from Washington to Baltimore.[22]

Thus, the sectors that could benefit most from the interactive capabilities of the telegraph did not become important users during its early development. Partly for this reason, perhaps, even as the system developed, telegraph managers did not realize its potential for two-way communications in real time. Once the system expanded, there was a rush of business over the wires. For efficient handling of the large number of messages, telegraph companies regulated the traffic by establishing rosters, giving each station a turn at transmitting a bundle of accumulated dispatches without interruption. Messages, rather than being transmitted on receipt, had to wait hours for transmission, making impossible "conversation," or "speaking over the wires," as the early enthusiastic newspaper descriptions had it.[23]

The telegraph was, therefore, especially well suited to users whose requirements were high-speed, one-way communications. A remarkably insightful promotional pamphlet, circulated by the entrepreneurs of the New York-Boston telegraph line, probably late in 1845, enumerated the groups that required such a service. The list was headed by bankers, who needed to have the "latest possible intelligence from the leading cities. The slightest variations in the money and stock markets has a bearing . . . upon their operations." Then came brokers, who similarly needed to "keep constantly advised of the state of the markets in neighboring cities"; and insurance companies. The government, police, travelers, and merchants were grouped as the source of "miscellaneous business."[24]

The list of main customers, those who must use "instant communication," also included newspaper editors. On them, however, the author of the pamphlet did not elaborate. "The necessity of their using the most rapid means of communication," stated the pamphlet, "is too obvious to require a passing remark."[25] The pamphlet also failed to point out that there was an important overlap in the demand for instantaneous communication by the press and the commercial sector. Long before the invention of the telegraph, the press had won its competition with the elite of the business community, becoming the leading medium for a monopoly of timely commercial news.

But with the appearance of the telegraph, the competition for news supremacy was back to where it had been in the days prior to the development of the express arrangements, when all communicators faced the same speed limitations. In the age of the telegraph there was no way to increase the speed of the standard device for transmission, and it was available to all communicators alike, at any time of the day or night. Private communications again had the temporal advantage over newspaper news, which appeared only at fixed intervals and became available only after typesetting and printing. Thus, with the advent of the telegraph, the press stood to lose its exclusivity in presenting the earliest news.

James Gordon Bennett, an acknowledged leader of the newspapers' news revolution, keenly realized the danger that the telegraph posed to the newspaper industry, and he initially was pessimistic about the fate of the press in the telegraphic age. "The telegraph may not affect magazine literature," surmised Bennett, "but the mere newspapers—the circulators of intelligence merely—must submit to destiny, and go out of existence." The *Alexandria Gazette* similarly assumed that since, with the telegraph, the public would be "furnished promptly . . . with the great events of the day," the press would have to limit itself to "examining causes, tracing effects, enlightening the judgments, and directing the reflections of men." In short, it would have to limit itself to an editorial function.[26]

It may have been that Bennett opposed the telegraph and initially tried to hinder its diffusion;[27] yet shortly he joined in his colleagues' enthusiasm for the invention.[28] Bennett came to realize that the telegraph offered the press great advantages, which offset its dangers. The attraction of speedy news was not necessarily confined to the quest for practical advantages to be gained from timely knowledge of economic or commercial developments. Editors found, at least since the advent of the penny press, that timeliness was an attractive attribute of social and political news as well.[29]

The telegraph, by increasing the speed of news and making its continuous transmission possible, broke down the reporting of developing news stories into smaller and more frequent segments. By rushing each new development to the public, editors enhanced the public's suspense as to the final outcome of the stories they were covering. This building up of expectations—what the media sociologist Helen MacGill Hughes called the "quickening urgency" of news—attracted readers and encouraged the sale of newspapers.[30] Thus Bennett ultimately predicted that with the telegraph, "the public mind will be stimulated to greater activity by the rapid circulation of news. The swift communication of tidings of great events will awake in the masses of the community still keener interest in public affairs."[31]

The advent of the telegraph also promised to expand the scope of the news. Now it would be news from all over the country, not merely local events, that could capture the attention of the public and create expectations as to future developments and the resolution of events. "The events of yesterday throughout the entire land will be given, as we now give the occurrences at home, to-day," anticipated the *Philadelphia North American* late in 1845. Obviously pleased, the *North American* predicted that by "feeding public curiosity," and thus increasing "the appetite for news," the telegraph would "revolutionize the press. Oriental imagination never ascribed to magic the power which this discovery will impart to it."[32]

Discerning editors also realized that by closing the time gap between the occurrence of exciting events and public knowledge of them, the telegraph enhanced their readers' emotional gratification. Following the development of distant events as they occurred, readers throughout the country shared the experience in time. They became, in a sense, participants in even the most remote news events. And this news, widely shared, satisfied the readers' craving "to be connected."[33] The experience of the penny press in speeding the news apparently made Bennett recognize that timeliness could broaden this exciting notion of common experience. With telegraphic news, he wrote, "the whole nation is impressed with the same idea at the same moment. One feeling and one impulse are thus created and maintained from the center of land to its uttermost extremities."[34]

To realize the great potential of the telegraph for advancing the press it was first necessary to push the wires beyond their dead end at Baltimore and extend them throughout the country. To Morse's disappointment, it became apparent early on that the government could not be counted on to undertake the project. Because neither he nor his close associates had the means for such an undertaking, the future of the telegraph came to depend on the goodwill of private investors.

Morse and Vail were wholly unqualified to promote their invention and to manage its large-scale diffusion. And they knew it, too. "I am not a businessman," lamented Morse, "I feel like a child in business matters. I can invent and perfect the invention and demonstrate its uses and practicality, but 'further the deponent saith not.'"[35] Morse, therefore, enlisted the support of three men of affairs to advance the cause of telegraphy and of his own interest in it. Amos Kendall, Francis O. J. Smith, and Henry O'Rielly became the apostles sent south, north, and west to implement the invention. All three men had been powerful players in journalism earlier in their careers.

As editor of the Frankfort (Kentucky) *Argus,* Kendall stirred a pot so lively that he was welcomed into the innermost councils of President Jackson. As a member of Jackson's kitchen cabinet, Kendall drafted many of the resounding state papers that had a strong impact on public opinion, and this ultimately landed him the job of postmaster general.[36] In this position, he was officially entrusted both with managing the operations of the newspaper exchanges that formed the backbone of American news gathering and with distributing newspapers to their distant readers. During his term of office, Kendall initiated express mail for the benefit of newspaper editors. His management of Morse's affairs, if not imaginative, gave the enterprise a measure of respectability it sorely needed.[37]

Less righteous, sometimes shady, yet ever colorful, was Francis O. J. Smith—newspaper proprietor, information manipulator, and telegraph promoter. Early in his career he secured an interest in several of Maine's leading newspapers. From this position he emerged as a key figure in propaganda campaigns of the Tyler administration intended to "adjust the tone and the direction of the party presses, and through them the public sentiment" toward support of Tyler and Webster's foreign policy, especially regarding the Maine boundary dispute. Smith was generously sustained by the president's secret fund in carrying out his covert missions.[38] His operations in the telegraph industry always seemed to bring him into conflict with both colleagues and customers. Smith was "a litigious man," concluded the R. G. Dun analyst who summarized his business career in 1859.[39]

Just as colorful a figure, but far more naive, was Henry O'Rielly, the romantic among the three. O'Rielly prided himself on having edited the *Rochester Daily Advertiser,* "the first newspaper between the Hudson and the Pacific," from its establishment in 1826.[40] He was an old hand at using the methods of what Daniel J. Boorstin called the "booster press" of the developing West.[41] O'Rielly, it turned out, took his own propaganda campaigns on behalf of the telegraph far too seriously. He overextended himself in wonderfully imaginative and ambitious projects and barely managed to remain solvent when his grandiose schemes crumbled. The least sophisticated of the three, O'Rielly appears to have done the best job promoting the telegraph in its first few years.[42]

All three apostles were well attuned to the requirements of the press, familiar with the modus operandi of newspaper editors, and attentive to their needs. Each had an intimate knowledge of the power of the press in promoting causes, gained from the experiences of their previous careers. These experiences, however, were varied, and they lent to each a unique approach to the "selling" of the telegraph. None of the men, however, found the task of promoting the telegraph an easy one.

From the start, it was apparent that investment in the telegraph would attract only risk capital. Shares in prospective telegraph companies were therefore allotted under a plan that had future success already built in. The companies that formed to establish telegraph lines were capitalized at three to four times the amount of cash deemed necessary for construction. Investors received shares representing 50 to 100 percent more capital than they actually invested. The patent rights were usually granted to promoters in lieu of half the stock of the new companies, rather than for cash. Thus Morse, Smith, Alfred Vail and his brother George, and Leonard Gale (a scientist who assisted Morse in developing the invention)—those who held the patent rights—would also have their financial gratification delayed until their patent proved it could make money for those who had invested in it.[43]

Kendall, who on 10 March was made the agent of Morse, the Vails, and Gale and thus represented three-quarters of the patentee interests, had more than future profits in mind when he framed his master plan for spreading the telegraph. By retaining half the voting stock of every telegraph company in the hands of the patentees, he believed it possible to promote unified policies and procedures that would bind the separate lines into a coherent national network. A flaw inherent in the scheme, which Kendall did not foresee, was the fragmentation of patentee ownership. As it turned out, disagreement among the patentees made concerted action impossible and ultimately threw the emerging industry into chaos. The haste with which the major contracts granting patent rights to entrepreneurs were drawn, most notably the contract with O'Rielly, left great ambiguities that ultimately frustrated hopes for the emergence of a coordinated confederation of telegraph companies.

State legislatures were all generous in granting acts of incorporation to the pioneer telegraph companies, allowing them much leeway. Consequently, none of the problems that hindered the integration of lines into a national system was due to conflicting state policies. Moreover, perhaps because so little was known about the actual costs involved in constructing telegraph lines, no legislature raised the issue of the inflated capitalization of the companies. And none demanded in return for the privileges it granted more than priority in the use of the wires in public emergencies or for the arrest of fugitive criminals. All state charters granted the right to build lines along public roads and canals and prescribed detailed penalties for damaging telegraphic equipment.[44]

Notwithstanding the fact that investors could receive double the stock they paid for, could use Morse's patent without paying for the privilege in cash, and could operate under favorable state laws, finding investors turned out to be a formidable task. Amos Kendall, seeking investors for the New York-Washington connection, directed his efforts toward enlisting capitalists from the commercial community of the major urban centers in the Northeast. "Not a sixpence" was raised in New York; nor did he fare any better among the businessmen of Philadelphia.[45] Kendall had some success in Washington and Baltimore. The first substantial subscription to the Magnetic Telegraph Company, formed to establish the telegraphic connection between Washington and New York, came from William Corcoran of the eminent banking firm of Corcoran and Riggs. The firm operated mainly on the Baltimore-Washington axis, and it used the pioneer telegraph line extensively.[46]

Few capitalists in Washington joined Corcoran in investing in the line. It would never have extended northward had newspaper proprietors from Baltimore and Philadelphia not come to the rescue. The largest investor in the Magnetic Telegraph Company was William M. Swain, the proprietor of the *Philadelphia Public Ledger.*[47] Swain, who subsequently became president of the company, also held stock in the name of Joseph Chandler, editor of the *United States Gazette.*[48] Other sums were advanced by Park Benjamin, the former editor of the *New York Signal* and current editor of the *Western Continent* in Baltimore.[49] Hart, of the *United States Gazette,* also purchased stock in the company,[50] as did several other editors converted to the cause.[51]

In Boston, Francis O. J. Smith fared no better than had Kendall in New York. He could not raise a penny for the projected Boston-New York line. In New York he did succeed in disposing of some of his stock; it appears that all takers were newspaper proprietors. The most substantial sum apparently came from Horace Greeley, a warm supporter of the invention from its infancy.[52] Bennett of the *Herald* also invested in the line,[53] as did James Watson Webb of the *Courier and Enquirer,*[54] and Moses Y. Beach of the *Sun,* an enthusiast of the telegraph since 1838.[55]

Smith never disclosed the fact that newspaper proprietors invested in his line; instead, he made it appear as if he had built the enterprise entirely from his own resources. The secretive nature of the investments by the press may well have been due to some unpublicized understanding of their future privileges in the transmission of foreign news. But

Smith was not interested in surrendering much additional stock in his line. He held on to large blocks of shares in many of the lines he built, even when investors could easily be found. By doing so, he was able to maintain absolute control without ever having to consult with, or to explain his actions to, other interested parties.

O'Rielly's policy was the opposite of Smith's. He unloaded his stock on whoever would take it, immediately putting the sums he raised to use in expanding his system. His strategy of raising funds for the telegraph was also in sharp contrast to Kendall's. He looked not to the circumscribed circle of potential investors with large capital but rather to the more modest—what some would consider "Jacksonian"—fortunes. An old Democrat and frontiersman, he looked for investors among the farmers, the small county traders, the town shopkeepers, each with only some tens or hundreds of dollars to spare for investment. O'Rielly did not present telegraph shares as opportunities to employ risk capital. Rather, he solicited townsmen to advance the immediate interests of their own hamlet and ensure its prosperous future by having it linked to "the great highway of thought."[56]

O'Rielly, though instrumental in Kendall and Smith's efforts to raise large amounts of money from the affluent presses of the East, could not use the same approach in up-state New York or in the Mississippi Valley. The western presses, small and short on budgets, could not be looked to as sources of major financial contributions. They were, however, indispensable to O'Rielly's attempts to reach the hundreds of small investors he needed. The press could spread the promise of the telegraph among the masses. If newspapers themselves could not invest, they could appeal to the public and boost the prospects of the new invention.

They did precisely that. Newspapers competed in grandiose descriptions of the benefits of the telegraph and occasionally adopted a very lofty tone. "The citizens of Brunswick in upper Missouri," wrote the *St. Louis Daily Union,* "have held a meeting to procure the extension of the telegraph to that place." The newspaper pronounced that support of the enterprise would do no less than sustain national unity. The telegraph, "the harbinger of newer and closer connection," which made "the most distant regions of our Republic . . . neighbors . . . should remove all doubts as to the safe extension of our National boundaries."[57] The *St. Joseph Gazette* played on the sense of isolation of the frontiersman: "What a glorious thought it will be for the Pioneer, standing upon the verge of frontier civilization, to exchange sentiments . . . with the facility of lightening . . . from the humble hatch of his rude cabin, with the citizens of the capital."[58]

The western newspapers of course had a direct and important interest in the development of the telegraph. Just like the press of New York, Baltimore, or any other point, they desired to bring speedy news to the public. They could thus be expected to be initially cooperative and helpful. O'Rielly, however, did not rely exclusively on the power of common interests to ensure the goodwill of the press. He also supplied them with much information about the telegraph, in the form of pamphlets, circulars, and clippings from the leading newspapers of the East. Many annotations that he added to his treasury of clippings from western papers indicate that he frequently helped editors by writing articles for them about the telegraph. Besides, he supplied newspapers in whose services he was interested with more tangible benefits: he spent large sums of money on advertisement in the press, which had the double function of promoting the telegraph and providing subsidies for cooperative editors.

In addition, O'Rielly gratified helpful editors by handing them shares in his lines. He later explained the practice as a way "to show my thanks to those editors for taking hold and helping me to raise money to build the line" at a time "when myself and the enterprise were little known to and had no claims on their columns." The shares were not bribes, he argued; they were justifiable since the gentlemanly editors involved "aided me in the enterprise long before they could have imagined that it would be in my power to requite their liberal aid."[59] His strategy, unlike the stock honoraria he lavishly distributed, paid bountiful dividends. "O'Rielly is the patron of the newspapers," acknowledged one competitor, who learned to his chagrin that O'Rielly's "liberal expenditure of money has its effects." O'Rielly could always depend on the support of the press in the battles with his competitors that emerged following his initial success.[60]

Kendall disapproved of the practices of his two colleagues. On one occasion he was accused of allowing a number of newspapers to send their messages "first in order, and thus throw into the hands of a favored few a powerful engine of monopoly, which will be more dangerous to our commercial interests than can well be imagined." Kendall vehemently denied the charge. In an unmistakable allusion to O'Rielly, he declared that he never had "conciliated the support or warded off the attacks of newspapers by presents or promises." Kendall added, "I have inflexibly refused all offers of profit or support to the telegraph lines or myself as an inducement to give one newspaper advantages over another." Here, he was apparently referring to Smith.[61]

On its second day of operations, the government telegraph line began supplying congressional news to Baltimore's leading evening newspaper, the *Baltimore Patriot.* It continued this practice daily, free of charge.[62] As the presidential elections of 1844 approached, the government telegraph arranged for expresses to bring returns from the South and the West to Washington, then to transmit them north for the benefit of the press. The undertaking was Morse's initiative, and it was carried out with the support of the postmaster general. The efforts and expenses proved worthwhile. "Our express beat Adams and Co.," signaled the elated telegrapher Henry J. Rogers from Baltimore. "After coming here it went to newspaper offices and then beat Adams and Co." The telegraphic response from Washington was short but expressive: "Ha ha ha V[ail]."[63]

The gratuitous use of the telegraph by the press was not unusual, since the government telegraph was initially free to all users. That Morse, Vail, and Rogers went to the trouble of collecting the news for the press was due, aside from promotional considerations, to the long-standing policy of government support of the public press. The same policy that dictated free transmission of exchange newspapers and slips in the mails made transmission of news by wire a government service.

When on 1 April 1845 the government line began charging the public tolls, free service to the press was discontinued. Consequently, interested newspapers had to replace the obliging operators with their own reporters. Up to this point the telegraph operators had been able to serve all the newspapers with one report; now several reporters had to prepare and send messages to the various papers. The wastefulness of such a situation was apparent, and soon correspondents appeared who undertook to supply several papers with news accounts, instead of each having its own representative at the Capitol. By May 1845, John Wills of Baltimore was sending Baltimore news to the *National Intelligencer*, the *Charleston Evening News*, and the *Alexandria Gazette*. For years he served as the agent for a large number of papers, mostly southern. This practice was duplicated on new telegraph lines as soon as they were established. Telegraph correspondents were not members of the editorial staff of a particular newspaper; they served whichever papers paid them. Nor were they necessarily committed to the newspaper press—they could transmit news to any customer interested in telegraphic news.[64] Thus, late in 1845, Glen Peebles published an advertisement introducing a service "for furnishing the newspaper press at distant points, commercial gentlemen and others desiring regular and special intelligence or private correspondence through the agency of the telegraph."[65]

This new vocation soon acquired a name: telegraphic reporting, and its practitioners were designated reporters.[66] The economic rationale of this new branch of business was simple. Reporters could satisfy numerous paying customers by one report, at the price of one transmission. They set their price below what it would have cost each customer to collect and transmit the news individually, and therefore, their proposition was widely attractive.

The saving to the newspaper and the profit to the reporter came at the expense of the telegraph company. O'Rielly, Smith, and Kendall each responded to this new contingency differently. The three were united in favoring preferential treatment of the press over private customers. Still in the process of establishing their lines, selling shares, and battling for rights of way over public and private routes, all three entrepreneurs needed a "good press." They also recognized that the press was potentially a large—probably the largest—customer of the telegraph. Publishing daily, the newspapers were also regular consumers requiring bulk service. Because they were themselves journalists, Kendall, O'Rielly, and Smith knew something about the national thirst for news and could anticipate the effect of the telegraph on the newspaper-reading public. They also knew of the extravagant measures editors were prepared to take to satisfy the public's demand for timely news. By setting press rates low and encouraging the use of the telegraph for news transmission, they could expect to whet the appetite of readers and editors for more telegraphic news. The press would thus become not only a large, regular customer but one whose consumption would steadily increase.

Here agreement among the three ended. Kendall's response to the emergence of news reporting, which diminished his line's revenue from press customers, was the most straightforward. Alternately criticized as inflexible and praised as righteous, he was as much a bureaucrat as a Jacksonian could be. There were no company regulations or moral grounds upon which he could combat the reporters or prevent their practices. He let them have their way. Smith was not so complacent. He simply dictated that every newspaper publishing a telegraphic dispatch that had passed over his wire, even if it was shared, would pay. The actual sender was charged full press rates, and additional users half price. Although some of the press complained, news reporters had very little leeway on Smith's Boston line once it was completed, in the summer of 1846.[67]

O'Rielly's western lines and the New York, Albany, and Buffalo line had a more imaginative solution. The companies themselves assumed the function of telegraphic news supply to the press of the regions traversed by their lines. The segment of the New York State line between Albany and Buffalo, completed early in June 1846, was inaugurated by a telegraphic conversation between the president of the line and editors along its route. The subject was the transmission of telegraphic news to the press.[68] In consequence of that conversation, the telegraph company began to supply the papers along its routes—which associated for that purpose—with a consolidated, uniform, telegraphic news report. It was far cheaper for each newspaper to receive such a report than to pay independently for its own news. Yet the telegraph company collected tolls from each newspaper, and its revenue from the single transmission of news yielded an income many times greater than any comparable message.[69]

Ultimately, the New York State line and the lines in O'Rielly's network were content to reap the profits from the policy of requiring all newspapers to pay the company for the one transmission. After initial attempts to have company operators collect the news and compile the telegraphic report, the northern and western lines retired from the field and left it to the professional news reporters, who contracted with local newspapers for collecting and compiling the news. The telegraph companies remained the recipients of generous payments from each newspaper for transmitting the news.[70]

For newspaper editors, however, combined receipt of a uniform news report presented problems. One was the elimination of competition for news beats and of individuality in reporting the news. Another problem was that the news reporters did not serve the press exclusively; they also supplied information to private operators. As a result, the press could no longer dominate the news field.

For the western press, the advantages of the wholesale arrangement clearly outweighed its shortcomings. Their major competitors were not other local newspapers but the New York press. They had to have telegraphic news to compete with metropolitan newspapers reaching local subscribers by the same mails that brought the local editors their exchange papers. The decline of individuality was a small price to pay for that advantage. Moreover, the western line, the New York State line, and the southern line passed through sparsely populated regions, where there was not much competition among the country presses. Characteristically, these small settlements could not sustain more than one or two local papers. Because there was not much competition in the first place, its further decline with the introduction of uniform reports was not very significant.[71]

Conditions were very different in Boston, Philadelphia, and particularly New York. Affluent newspapers of the major cities could afford their own news reporters and their own telegraph bills. What they could hardly afford was the disappearance of competition among themselves and, even more important, competition with alternative sources of information. The giants of New York journalism were not intimidated by telegraph tolls for messages from Philadelphia, Harrisburg, or Albany. It would take much more to make them give up the bliss of a news beat or to surrender their supremacy in timely news to "speculators" from the private sector.

3

"A Better-Organized System": Establishing the New York Associated Press

The metropolitan press of the mid-nineteenth century promoted the gospel of individualism.[1] The newspaper itself was a highly personal enterprise. Proprietor-editors dominated the management, structure, and policy of their papers, and competition between dailies took the form of direct and aggressive rivalry between individuals. Newspapers then dedicated much space to vociferous attacks on their counterparts, and occasionally these wars of words escalated into street brawls between rival editors. The penny press was supposed to reflect the aspirations of the enterprising individual aroused by open and limitless opportunity. Against this backdrop, editors tended to emphasize the uniqueness of their papers, to de-emphasize common features, and to avoid cooperation.

Moreover, the popular press flourished on the egalitarian, anti-monopoly sentiment characteristic of the Jacksonian era. The leading newspapers commonly depicted themselves, in the polemics of laissez-faire, as independent organs striving to break monopolies of knowledge held by the privileged. In such an atmosphere, a combination intended to control such a sensitive resource as news could hardly generate popularity.

Yet the structure that ultimately emerged for telegraphic news gathering was one of cooperation. An association of leading New York editors, commonly known as the New York Associated Press, became the most significant institutional response of the American press to the novel communication technology.

Elucidating the causes underlying the formation of the first major American news wire service is essential for understanding the history of news communication in the telegraphic era. Establishing the date of its formation has been the subject of much interest and considerable disagreement.[2] Currently, May 1848 is generally accepted as the date of the founding of the Associated Press, but there are persistent doubts as to its accuracy.[3] A reliable dating of the establishment of the Associated Press will identify the immediate background from which it emerged and thus provide a better understanding of its causes and functions.

The most popular daily in the New York City of the mid-1840's was the *Sun*. Perhaps, as its owners claimed, it might even have been at the time the largest-selling newspaper in the world.[4] The paper was cheap, small, and it was energetically and compactly managed. Established in 1833, the *Sun* was the first successful New York penny paper. Sold in 1838 by its founder, Benjamin H. Day, to his brother-in-law Moses Yale Beach, it was run as a family affair. The Beach boys, Moses Sperry and Alfred Ely, together with their father, actively managed the paper; and in 1848 ownership of the giant passed to the two sons in partnership.

There was good reason the *New York Sun* was number one in circulation. Under Day's management, and especially after the mechanically talented Beaches took command, the *Sun* was innovative in adapting the most advanced technologies to newspaper production. It was among the first American dailies to use a steam-driven press, and together with Robert Hoe and his company, it led the way in advancing the newspaper printing process, thus shortening the interval between acquiring news and distributing it through its mammoth editions. Beach even contributed to improving the production of paper.

The *Sun* also excelled in adapting new technologies, as well as perfecting established arrangements, to accelerate the speed with which news reached its presses. Chartering fast locomotives, dispatching carrier pigeons, even setting special editions in type on board express steamships that carried late news were characteristic of the newspaper's efforts. The public, Beach knew, wanted the latest news, and leadership in New York journalism implied success in the frenzied race for it. In that context, the *Sun* advertised itself as "the most enterprising of its neighbors."[5]

The *Sun* could hardly have been expected to relinquish its technological leadership and its orientation to experiment with new devices once Morse introduced his revolutionary communication technology. Indeed, Beach became one of Morse's earliest enthusiasts and warmest supporters. After all, his newspaper prided itself on having "always been in advance of its contemporaries in rapidity of action for supplying public wants . . . In the means taken for furnishing its readers with the earliest news," boasted the 6 March 1848 issue, "the *Sun* has always led the way."

Surprisingly, perhaps, the self-proclaimed city leader in news enterprise was also responsible for bringing about cooperation in news gathering among the competing New York dailies, and this was just as the telegraph was developing into a powerful new force in the news business. The *Sun* engineered the movement toward cooperation just as it was scoring one of its most decisive news-gathering victories, a potentially devastating blow to its rivals. The occasion was the contest for Mexican War news.

"When the Mexican war broke out [in April 1846] there was, of course, great competition for the earliest news," reminisced retired Moses Sperry Beach in an interview in 1884 with a reporter for the *New York World*.[6] He was sent south to expedite the transmission of the war news between New Orleans, where the news first arrived by steamer, and the southern terminus of the telegraph line leading to New York.[7] Young Beach succeeded in arranging for a daily express between Mobile and Montgomery, which, when successful, saved at least twenty-four hours.[8] The arrangement stipulated the payment of some $700 for every successful express. On 29 May 1846, the *Sun* published the first news reports expedited by the Mobile-Montgomery express.[9]

Returning from his mission, "as pleased as a dog with two tails at [his] success, and at having made an arrangement which would always insure a beat of twenty-four hours for the *Sun*," Beach was summoned into his father's editorial sanctum. There, he found the editors of the other leading New York City papers. He recounts:

> My heart went down clear into my boots, but my father said, "You may tell these gentlemen any arrangement you have made; I have told them that such of them as like to go in and share the expense can do so." This was the beginning of the Associated Press. It all grew out of this. The *Courier and Enquirer*, the *Journal of Commerce* and *Express* were the three large papers [in format], and the *Sun*, *Herald* and *Tribune*[10] the three small ones—six in all.[11]

That Beach's account was not the idle bragging of a retired newspaper legend to a young and admiring colleague is established by a memorandum he wrote, apparently for his own records, on the establishment of the New York Associated Press.[12] The memorandum confirms the main features of the interview with the *World* reporter, including the dating of the beginning of the Mobile-Montgomery express to 1846. It also provides the list of persons present at the meeting and the order in which the other leading New York dailies accepted the *Sun*'s proposal of a shared service for procuring Mexican War news.

In the spring of 1853, Beach, as the association's secretary, presented a similar version of the origins of the Associated Press in the pages of the *Sun,* which he then owned in partnership with his brother.

> The present Association of the New York Newspapers was first started by the present proprietor of the *Sun* on the occasion of making arrangements for a daily express to secure the transmission of news from the South, and particularly from the seat of War in Mexico, in advance of all ordinary channels of communication. The objects and aims have been ever in accordance with the best interests of the whole community, whose confidence it had deservedly received and constantly merited.

This statement was published not many years after the event and was quoted in a widely circulated polemic pamphlet.[13] Although others subsequently claimed credit for establishing the New York Associated Press, no known refutation of this statement came from any quarter when it was published.[14] The *Sun*'s leadership, and Beach's standing as the "generalissimo" of the New York dailies in their cooperative efforts, were well enough recognized at the time.[15] Moreover, Beach's emphasis on the role of the Mexican War news arrangement in the establishment of the New York Associated Press is positively supported by other, well-informed contemporaries.[16] Yet the particular relation between running the express and cooperative arrangements for joint telegraphic transmission remains moot. Beach simply asserted that "the original Telegraph News Association . . . was based upon a contract for a daily express," but he did not elaborate on the provisions for transmitting the express news to New York.

The agreement to cooperate in procuring Mexican War news, reached in 1846, was, however, only one step in the emergence of the Associated Press as the major telegraphic news-gathering mechanism of New York City's leading dailies. Further agreements for cooperation between the dailies followed quickly. These included arrangements for procuring telegraphic news from Washington, Baltimore, and Boston, as well as shipping news from the port of New York, in common. In all of these, the *New York Sun* figured prominently. The Mexican War news agreement was apparently closely connected with, perhaps even contingent upon, the first of these additional cooperative efforts—the Washington news arrangement. This arrangement provided for telegraphic transmission of news reports from points south to Philadelphia and New York.

At the time, New York was being brought into direct telegraphic communication with Washington. First, the New York-Philadelphia segment was completed, and then, on 5 June 1846, Baltimore and Philadelphia were telegraphically linked. Over these new routes, connecting in Baltimore with Morse's pioneer line, messages could pass through from the nation's capital to its great metropolis. From Baltimore, news carried by the southern mail could be telegraphically transmitted to New York. Setting out to employ this "great highway of news," New York's editors found a Philadelphia news connection extremely useful.

Philadelphia's dominant dailies, soon to be dubbed "the Unholy Alliance," responded to the advance of the Magnetic Telegraph Company's line to Baltimore by forming an association to jointly receive Washington reports and news by the southern mail transmitted telegraphically from Baltimore. They were led by the *Public Ledger*'s William Swain, who also happened to be a large shareholder and director of Magnetic. Local agencies connected with the Philadelphia association made the news received by the Philadelphia editors available to New York newspapers. To receive the most current news from the South and from the capital, the New York papers had only to arrange—and pay for—the telegraphic transmission of the Philadelphia association's news from Philadelphia to New York.

Only seven of New York's twelve dailies responded to the challenge. The penny *Sun, Herald,* and *Tribune* combined to receive a joint report, and the six-penny *Journal of Commerce, Courier and Enquirer,* and *Express* combined to receive another. The *True Sun* paid for and received the Philadelphia report on its own. This alignment was, however, only temporary. Moses Yale Beach took command and, through a "process of negociation [*sic*]" with his local counterparts and the Philadelphia association, engineered a new, comprehensive arrangement for telegraphic news reports that took effect no later than 22 June 1846.[17]

Through Beach's initiative, the six leading New York dailies—the *Sun, Herald, Tribune, Journal of Commerce, Courier and Enquirer,* and the *Express*—associated to receive a common telegraphic news dispatch over the Magnetic line and divide its cost evenly. Together they covered the tolls for the Philadelphia-New York leg and shared the expense of transmitting the Washington and Baltimore news to Philadelphia with Swain's newspaper association. According to Magnetic's president, Amos Kendall, the reports from Washington were "sent for Mr. Beach of the *Sun,*" and a copy of these dispatches was apparently dropped at the Philadelphia telegraph office, for the local association, in the process of transmission to New York.[18] Five other New York dailies may have been approached, but they declined to share in the benefits of Beach's association.[19] One daily, however, was barred from the new group, notwithstanding its eagerness, even its desperate attempts, to join. This was the *New York True Sun,* established and owned by former employees of the *Sun,* its founder, Benjamin Day, prominent among them.

Beach would not have his brother-in-law, former partner, then competitor, and finally bitter enemy join in the benefits of the combination he had arranged in New York. Beach also prevented his brother-in-law from receiving the association's news in the City of Brotherly Love. On 22 June 1846 the *True Sun* was notified by its erstwhile Philadelphia news agent that he was "compelled to discontinue" the service "in consequence of Mr. Beach having agreed with the Company of Publishers here [in Philadelphia] to have the exclusive use of the news received for them at this point by paying a share [of their] expenses. The arrangement of Mr. Beach is for six New York papers, excluding yours."[20]

The *True Sun* would now have to procure congressional reports from Washington and southern news from Baltimore on its own. It would pay five dollars for the same quantity of news that cost each of the New York associates forty cents. Day and the *True Sun* decided to bear the burden, however uncheerfully. The paper heaped columns of abuse on Beach, on Swain and his Philadelphia association, on the telegraph monopoly, and on the news monopoly it had facilitated.[21] Nor did the *True Sun* spare the members of Beach's new "par nobile fratrum," those "lately warring antagonists but now loving co-laborers and sworn friends," the five New York dailies that took the "ignoble" role of "contended tools of Beach's enmity."[22]

The *New York Express*, in a mild response to these accusations, denied any intention of injuring the *True Sun*. It stated that for its part it would have been perfectly willing to admit Day's paper to what it termed "the association organized to receive news." The *True Sun* rejected the *Express*'s apology but, quite significantly, adopted its terminology. It now referred to Beach's newspaper combination as the "associated press of this city."[23] This was the term that ultimately came into general use in reference both to the six individual dailies that engaged in cooperative news gathering and to the common framework they organized.[24]

The New York newspapers that were now joined together for rapid news gathering from Washington, and for its joint telegraphic transmission, were apparently well satisfied with their cooperative arrangement. The records of the Magnetic Telegraph Company indicate that the association continued long after the summer of 1846. Thus, on 24 November 1847, the board of directors of the Magnetic held a special meeting to consider "a proposition made by the newspaper press of New York, Philadelphia, Baltimore and Washington" concerning new rates for the transmission of news reports. The papers' proposition pertained to news that they received "collectively," through the arrangement "by which they all [could] receive the Congressional Proceedings and general news by a single transmission."[25] Surveys of the actual copy published in the leading New York newspapers also demonstrated that the papers published uniform wire reports from Washington from the onset of through telegraphic communications between the two cities, and ever after.[26]

Developments in the Northeast placed additional pressure on New York papers to combine. Here was another bottleneck of information transmission, which involved the flow of European news. In 1846, Samuel Cunard inaugurated a fleet of swift ocean-going steam vessels that made regularly scheduled Atlantic crossings. Cunard steamers could be depended on to bring the latest European news to the American continent. No ship departing Europe after a Cunarder was speedy enough to overtake one of these swift ocean steamers on the westbound voyage. The regularity of sailing schedules and the ships' relative independence from the vagaries of weather conditions promised a high rate of on-schedule or close-to-schedule landings. Samuel Cunard, an obstinate man, insisted on keeping Boston the American terminus for his line. Thus, to the distress of New York editors, the latest European news on board the Cunard fleet of 1846–1847 had to be collected at Boston Harbor.

Late in June 1846, the inauguration of telegraph service between Boston and New York promised some help to the New York newspapers in their effort to speed steamer news to their readers. According to Moses S. Beach's account, however, it appears that the remedy induced maladies of its own. The New York dailies, he recalled, sent representatives to the telegraph office "twenty-four hours ahead" of the expected arrival of steamers, "and [they] stayed there night and day." Henry J. Raymond of the *Courier and Enquirer,* Andrews of the *Journal of Commerce,* Hudson for the *Herald,* together with Moses S. Beach for the *Sun,* found that for all the inconvenience, they were only duplicating one another's news. "We, for our own convenience," recalled Beach, "finally decided to take turns to receive and manifold the news . . . This arrangement finally merged as the Mexican war news had, into the associated papers."[27] Indeed, as early as 18 July 1846, steamer news reported in three major New York dailies was derived from a common source. Identical telegraphic reports thus began soon, but not immediately, after the opening of the New York-Boston telegraph line on 27 June 1846.[28]

Beach and his counterparts soon improved the arrangement by having the *Boston Daily Times,* a Beach property, procure the steamer news on their behalf. A representative of the New York papers, however, made up the telegraphic reports from the European newspapers and transmitted them to New York.[29] Thus, Moses Beach's brother, Alfred Ely Beach, testified that on 7 November 1846 he was in Boston "for the purpose of forwarding the news per steamer *Britannia* to the New York press generally . . . I went there as agent and correspondent for six or seven different daily newspapers."[30] If at first the associated dailies entrusted the *Sun* with the responsibility of sending the steamer dispatch, they soon found a better arrangement. They hired John Turel Smith, an established Boston news reporter, to act as their regular agent.[31]

Verification of Smith's employment can be found in a sworn statement presented to a committee of the Massachusetts legislature in 1859 by Daniel H. Craig, the general agent of the New York Associated Press, who recounted the history of the organization he managed: "*Ever since* the first line of telegraph was established between Boston and New York . . . more or less of the newspapers in both cities have been associated together in their telegraph news arrangements . . . My acquaintance with the New York Associated Press dates back to the winter of 1848–49 . . . Mr. John T. Smith was then, *and had been for several years before,* the Boston agent of the New York Associated Press."[32] The New York editors apparently also engaged an agent to receive the telegraphic dispatches of steamer news in New York. The identity of this "gentleman depending on the [Boston] line for the foreign news in behalf of the New York Press" is, however, unknown.[33]

But Craig's unequivocal "sworn statement of facts" appears to be contradicted by the one piece of evidence on which the establishment of 1848 as the date of the founding of the New York Associated Press depends. On 13 May 1848, Henry J. Raymond, at the time James Watson Webb's assistant on the *Courier and Enquirer,* addressed a letter to the potentate of the telegraph line connecting Boston and New York, F. O. J. Smith (made known throughout the country as "Fog" Smith by the less than admiring press of New York).[34] "The *Journal of Commerce, Express, Courier and Enquirer, Herald, Sun,* and *Tribune* of this city," wrote Raymond, "have agreed to procure foreign news by telegraph from Boston in common, and have appointed a committee to make arrangements with you for its transmission." It was on behalf of the committee that Raymond addressed Smith.

However, Smith's rapid response, dated two days later, undermines any thought that this exchange referred to the actual inauguration of common news gathering by the New York press in the eastern theater of operations. "Your letter," wrote Smith to Raymond, "proceeds on the supposition that your *existing arrangement* with the New York and Boston Telegraph is discontinued . . . This frequent *change of arrangement* is annoying to all parties, and hence, for the sake of permanency and responsibility, I will contract . . . for one year."[35] Rather than establishing the inception of the Associated Press in 1848, the Raymond-Smith correspondence provides further support for an early dating of the association's connection with the Boston line.[36]

Several circumstances made the formation of a wire service both necessary and possible in the summer of 1846. That summer witnessed, in rapid succession, the convergence of isolated telegraph lines in New York City. With the opening of the Philadelphia to Baltimore segment early in June, the New York and Washington trunk line was completed. Within a month Boston and New York were telegraphically connected. New York was linked to Albany by telegraph shortly afterward, in September 1846, and through Albany direct communication with Buffalo was made possible. Almost at once, previously isolated telegraph lines were transformed into a network, and New York found itself at "the center of this magnetic circle," emerging as the nerve center of an American telegraph system.[37]

Lines now spread east of the city to Boston, and north and west to Albany and Buffalo. The southern line to Philadelphia and Washington was in the process of extension southward along the seaboard, toward New Orleans. Via Philadelphia and Pittsburgh the telegraph was beginning its westward march, bringing communities in the Mississippi Valley into direct communication with the seaboard centers. New York became the "one grand center" of this network.[38]

Each of these routes was significant from the standpoint of American news flow. War news and congressional reports came from the South, steamer news from the East, and political reports from the state capital in the North. Although not yet fully integrated, the telegraph lines converging on New York were developing the capacity to communicate messages from one telegraph station to any other, be it in Rochester, New York, or Richmond, Virginia, in Lowell or in Louisville. News items from every part of the country traversed by telegraph lines could be transmitted directly to New York, and news accumulated there could be "broadcast" throughout the country. The New York press, for the first time, was employing a technological system and interacting through it with an extended field. The military historian Martin Van Creveld gives a clear summary of the implications of operating in what he calls "the age of systems":

> It was not so much that the newly invented tools and machines were more effective than their predecessors, though that was important, as the fact that they could only be used in integrated systems which did the trick. As more and more individual tools and machines surrendered to the drive for efficiency and were inte-

> grated into systems, success . . . tended to depend on nothing as much as the ability to understand these systems and manage them . . . The system approach [was] first made manifest in the telegraph and the railway.[39]

The telegraph was clearly more effective than its predecessors in transmitting news. When newspaper establishments began employing it in earnest, they faced a serious challenge in understanding and managing a system of telegraphic news gathering.[40]

Before this time, newspaper exchanges formed the skeleton of American news gathering, and the control of a continuous flow of news from a wide field was a relatively simple task for the individual editor. He established a set of bilateral exchange agreements with editors in other parts of the country. Once set up, this foundation was more or less permanent. The United States Post Office Department would then, through its own prearranged network of schedules, transmit the news through the maze of routes formed by the cumulation of these exchange agreements. Thus, once the foundations were laid, the system more or less managed itself.

A great price in wasted resources was paid for the organizational advantages that the exchange arrangement provided.[41] As early as 1794, Postmaster General Timothy Pickering was complaining of the wastefulness of the exchanges. "One of the Baltimore printers lately informed me that he make[s] no other use of most of them, than to furnish the neighboring shops with wastepaper."[42] The dead weight of usually unusable but occasionally indispensable exchange newspapers was carried at the expense of the post office. The newspapers did pay for paper and for the printing of a great number of unsalable copies. Moreover, the system also heavily taxed editorial time.[43] In the early nineteenth century, however, paper, printing, and editorial time were much cheaper and more abundant resources than speedy communication systems or organizational talent.

News gathering by telegraph differed dramatically in its organizational implications from the editorial exchange. It transformed news gathering into a dynamic process, requiring active control of a complex, interactive, and potentially continuous operation. Very much like the expresses run by the press, and very much unlike the exchange arrangement, telegraphing news cost money. The non-selective and redundant flow of news characteristic of the exchanges became impracticable when the government was not footing the bill for transmission.

With the telegraph, information was transmitted only when it was of value to the receiver. The selection, however, had to be made by someone in the field, actively seeking such information and remaining in contact with its intended receiver. An editor could now also call for news from any quarter he considered important, provided that he had arranged for a connection in that place and that he could specify to his news source the information he required.

In colorful verbiage, the *Baltimore Sun* tried to explain to its neighbor, the *Patriot,* precisely these requirements of active news gathering. This was in May 1846, after the *Patriot,* badly beaten on an important news item, blamed the telegraph. "The persevering blunder of the [*Patriot*]," suggested the *Sun,* "is attributable, very evidently, to an entire misconception of the nature of the Telegraph . . . An hallucination [led the *Patriot*] to discern some imaginary being allied therewith, which enjoys access to mails, newspapers, letters and all verbal information, a creature Argus-eyed—Briareus arm'd—Donkey-eared, we suppose, and Griffin-winged—a sort of nondescript animal born of the Telegraph itself." The sole duty of this creature, according to the *Patriot*'s misconception, was to furnish intelligence at one end of the wire, for the benefit of "those *newspapers* which deem it too much trouble to look after the *news* themselves" at its other end.[44]

The major New York papers were aware of the managerial implications of organizing the collection and transmission of news even before the advent of the telegraph. A highly eventful attempt by the *New York Tribune* to express steamer news from Halifax to New York, in the spring of 1846, illustrated the inability of the individual newspaper to handle complex operations. Competitors claimed that it took the *Tribune* express sixteen hours to complete the Portland-New York leg alone. "No such thing," answered the newspaper.

> We lost some fifteen minutes at the ferry on the east side of Boston. Then a very short time (instead of an hour and a half, as it is reported by the *Express*) in finding our agent in Boston; then an hour in firing up an engine and getting away from Boston, where all should have been ready for us, but was not. The locomotive was over two hours in making the run to Worcester—42 miles—[double the optimal time]. At Worcester, Mr. Twitchell (whom our agent on this end had only been able to find on Tuesday, having been kept two days on the route to Boston by a storm, and then finding Mr. Twitchell absent in New Hampshire) was found in bed, but got up and put off, intending to ride but one stage. At this end, however, he found the rider he had hired sick, and had to come along himself. At one stopping-place he found his horse amiss, and had to buy one before he could proceed.[45]

This comedy of management errors by the *Tribune,* when considered against the structure of the contemporary daily, is not surprising.

Newspapers at the time, though commercial giants, were primitive operations from the standpoint of organization. Legends about the *Tribune*'s editor, in the early phase of his journalistic career, illustrate this fact. Horace Greeley, a colorful figure, generated such an abundance of anecdotes that even his plain white coat became legendary. One coat story has it that Greeley, on the first cool day of fall, reached for his winter coat and found in its enormous pockets some two hundred unopened letters. It turned out that the absent-minded editor had picked up the batch of letters at the post office on the last cool day of spring, after which he dispensed with the coat for the ensuing warm season. The discovery of the letters helped settle the mystery of a barrage of complaints from subscribers who claimed pre-payment, suppliers who complained of non-payment, and other correspondents who charged discourtesy.[46] Greeley's eccentricities aside, this story offers an insight into the operation of a major newspaper in 1840. For it was the editor himself who went down to the post office to receive the mail; and he was also charged with updating the subscriber lists, settling bills with suppliers, and so on.

Whatever may have been meant by the occupation category "newsman" in the census of 1850, there were only 209 people so occupied to run the 2302 newspapers numbered in that census. Descriptions dating from the second quarter of the nineteenth century characteristically depict the country paper as a one- or two-man effort. Because the majority of those publications were adjuncts to printing establishments, the production of the paper was done in-house, usually with the aid of apprentices. The proprietor himself culled the news from the exchanges, wrote some paragraphs, arranged for occasional correspondence, and, after the printing, marketed, and sometimes even delivered the newspapers himself.[47]

Even a daily newspaper in the greatest metropolis did not require an elaborate organization in order to succeed. The *Sun* was established and operated by a partnership between a printer turned editor and a police reporter who employed two professional printers to produce the sheet. Such a team was sufficient to attract 2,000 readers within two months of its first issue (3 September 1833) and to establish a top-ranking city circulation of 8,000 copies within five months. James Gordon Bennett achieved the ultimate compactness of newspaper structure. When the *Herald* first appeared, on 6 May 1835, he carried out all the functions of his daily, aside from printing, singlehandedly. "I have written my own editorials," Bennett later recalled, "my own police reports . . . my own Wall Street reports . . . my own squibs, crackers and jeux d'esprit.—I have been my own clerk and accountant—posted my own books, made out my own bills and generally attended to all business details in the office." The *Herald* reached a circulation of some 15,000 before the newspaper doubled its personnel with the employment of George Houston late in 1835.[48]

By mid-century there developed a differentiation between the editorial and business functions of the larger newspapers, which brought about the creation of separate departments. Sub-divisions appeared within these departments, with subscription, advertising, and production emerging as distinct units of the business department. The editorial department remained somewhat more amorphous, and the differentiation of functions within it developed only gradually and was usually unique to each newspaper.[49] Even after these developments, however, the metropolitan newspaper remained a compact organization. The popular *Boston Transcript* as late as 1852 had a three-person editorial staff—a woman editor and two male reporters.[50] The penny *Philadelphia Public Ledger* in the early 1840's employed an editor, a news editor, and one reporter.[51] John Bigelow, joining the staff of the prestigious *New York Evening Post* in 1849, found there a chief editor, an exchange editor, a city reporter, and a marine reporter who also worked for the *Commercial Advertiser.* Bigelow claimed that "that was about the condition of all the newspapers published in New York at the time."[52]

The *Tribune* directory, ten years after the introduction of the telegraph, when the paper had a circulation of 30,000, did include a considerable number of editorial employees, some 150 of them. A closer look at the list indicates that more than 100 of this impressive staff were employed "in common with the Associated Press" or, more simply, were Associated Press agents. Of those who served the *Tribune* exclusively, 25 persons were employed in the editorial department, and 8 more in the "publication," or business, department.[53]

The compact and primitive nature of the newspaper's organization and management made integrating the elaborate structure involved in telegraphic news gathering impossible. The continuous coordination of a large and widely dispersed network interacting with a central headquarters through a system of complex communication machines had to be done, at least at first, out-of-house. With the transformation of American telegraphic lines into a system in the summer of 1846, it became patent that specialized and independent arrangements were required to reap the full benefits of Morse's invention. Thus, the nature of the telegraph as a system, coupled with the limitations of the managerial structure of the contemporary daily, brought about the cooperative agreements reached in rapid sequence in the summer of 1846. This was the underlying cause for the formation of the New York Associated Press.

Although the revolutionary implications of operating a technological system in real time may not have been fully obvious to contemporaries, they were aware of several more immediate incentives for cooperation in telegraphic news gathering. The high cost of telegraphic transmission has frequently been proposed by students of the wire services as the reason for the emergence of the Associated Press. Telegraphing news certainly did cost more than sending it through the mails, and dividing the bill made it cheaper. Yet for each of the affluent presses of New York City that combined to form the Associated Press, telegraph tolls were a pittance to pay for a scoop.

In fact, occupying the full yearly transmission capacity of the entire American telegraph system in 1846 would have cost a newspaper less than $50,000.[54] This sum would have represented a significant savings in the news transmission budget of each of the major New York dailies as compared with previous outlays.[55] Twenty years earlier, the *Journal of Commerce* and the *Courier and Enquirer,* with circulations of about 5,000 copies each, spent a similar amount yearly on their express news arrangements alone. Moreover, it would only be necessary for a New York daily to purchase half the capacity of the wire to monopolize all incoming information to New York, leaving the other half of potential transmission time for outbound messages. Controlling the entire national potential of telegraphic transmission and thereby monopolizing all important scoops for $25,000 was a bargain by the standards of the New York press. Thus, in the early days of the telegraph, price was hardly a consideration potent enough to force newspapers into cooperation.

The direct incentive for cooperation, a distasteful course for the competitive and aggressive dailies of New York, obviously lay elsewhere. Professor Morse, as superintendent of the government telegraph line between Washington and Baltimore, was harassed by disgruntled editors before he found it necessary to lay down the "first and essential principle of telegraphy"—that the first to come should be the first to be served. Although this appears to be a fair enough procedure, in full accord with previous conventions of the news beat by express, this rule alone would not suffice. Unlike the mailbag, the telegraph wire could transmit only one message at a time, and though the telegraphic signal was transmitted instantaneously, the sequence of characters composing the message had to be signaled one by one, by independent manipulation of the telegraph key. Consequently, much time was required for operating the revolutionary device for the "annihilation of time and space."

The characteristic stratagem employed to outsmart the competition under the first come first served rule was indeed to get there first. News reporters appeared in the telegraph office hours or even days in advance of anticipated information, equipped with the Good Book, and had the obliging operator commence sending their message with "In the beginning . . ." Daniel Craig testified that occasionally the operator approached the end of Deuteronomy before he was presented with a more newsy text for transmission.[56] The price of occupying the line for hours on end was inconsequential by the standards of the New York press.

While virtually every veteran news-gatherer later claimed to have resorted to the trick of occupying the wire while waiting for news, such claims should not be taken too seriously. The opportunities to monopolize the telegraph wire in this manner were extremely limited, as the rules of priority were soon modified. Editors and, more important, telegraph proprietors had good reasons to change the unbridled "first come first served" arrangement.

Some of the pressures that led to amending the rule were demonstrated when President Polk's message of 1848 was reported to the press. Tuesday afternoon at two o'clock the first words of the president's message began flashing over the Pittsburgh and Louisville telegraph line. The message, which related to the Mexican situation, was a landmark in American foreign policy, as well as in length of presidential pronouncements. The last words of the message were received in the cities of the Mississippi Valley some forty-eight hours later. Bad weather and "atmospheric interference" slowed the transmission, which had originally been expected to require twenty hours. There was, however, an additional significant delay. The passage of the message over the line was stopped for five hours on Wednesday afternoon by the superintendent of the line, so that other messages that had been accumulating in the telegraph offices along the line could be transmitted. The superintendent, James D. Reid, tried to explain the nature of these messages to the outraged press:

> A merchant in Cincinnati wanted to pay his note, due that day in New York. His dispatch must be delivered before 3 o'clock. At 1 [o'clock] it was on our files unsent. A mother was dying; she wished to see her boy—her eldest, her best beloved—before she yielded up the panting breath that struggled to be free. Her message was there, too. A merchant, his goods seized, the hammer of the auctioneer ready for the sacrifice, appealed to his distant friend for aid. His message is there unsent. A hundred other of various concern were thus grouped, the expressions of many hearts trembling, hoping, fearing the result of their distant mission. An hour or two, and these would have failed of their purpose . . . and for what? For a message justly interesting, but involving to none living an individual joy, or advantage, or regret.[57]

The interruption in the transmission of President Polk's message indicated to all that something was wrong. Reid, who was responsible for the delay, counted a net loss of two hundred dollars for his line as a result of sending the message. The press of the Mississippi Valley, for its part, raised a furor at the delay in the transmission of the message. After all, it had come first and was entitled to full, uninterrupted service. The true cause of the discontent was that the capacity of the telegraphic channel between Pittsburgh and the Midwest was simply not sufficient to satisfy the demand for its services.[58]

Having the press monopolize telegraph lines for long periods of time was at first a blessing for the fledgling telegraph industry. Without the strong support of the press, the lines would have remained idle throughout most of the day. But once the public began to patronize the telegraph in earnest, it became clear that there was not enough transmission time to go around. Moreover, telegraph companies stood to lose when enterprising news reporters dominated their lines with long and therefore only marginally remunerative messages. Much like contemporary taxi drivers, the early telegraph companies could make more money from many customers requiring a short service than from a few lengthy runs. Thus Alfred Vail, Morse's associate and superintendent of the government line between Washington and Baltimore, found in the summer of 1846 that putting the line at the exclusive service of the press for ten hours of uninterrupted transmission (which he considered the optimal workload for a single day) would yield $33.34. But a full day of telegraphic service for the public at large would garner $57.60, or an income 73 percent higher.[59]

Nor were newspaper editors satisfied with the prevailing arrangements. Even the most risk-loving editor faced the specter of expending a fortune on a news item and then losing the whole investment if his reporter reached the telegraph office seconds after a competitor. Then again, there was always the danger of a private interest monopolizing the wire to the exclusion of the press. Newspapers would be collectively beaten in their quest to control the flow of American news. "Under existing circumstances serious difficulties occur," argued the *Rochester Democrat* in 1847. "In consequence of there being but one or two wires connecting with the great cities, an individual who reaches the office first, and is anxious to transmit the contents of a long document, is able to monopolize the Telegraph."[60]

The inevitable solution was to alter the first come first served principle. The remedy came in the form of the fifteen-minute rule. This rule, wrote William Shanks, an active journalist of the time, "required that in the transmission of the telegrams to the press, the operator should send to one paper for a quarter of an hour, then to another for the same period, and so on until each was served in turn."[61] In this last detail, the writer was apparently affected by the egocentric perspective commonly attributed to members of the fourth estate. The fifteen-minute rule was not exclusive to journalists; it simply stated that "no individual or combination of individuals shall have the use of the telegraph for more than fifteen minutes at one time, when any other individual is waiting."[62]

Once established, this procedure for time sharing virtually nullified the advantages of primacy in news gathering. The first to reach the telegraph office had exclusive access to transmission facilities only for the actual time by which he beat his competitors. Considering the slow pace of telegraphic transmission, such time margins were characteristically not sufficient to have the full message at the home editorial office ahead of the competition to ensure exclusive publication. In these circumstances, there was not much point to competition. Because all newspapers ended up more or less on the same footing, the energy and outlays expended on individual efforts were not remunerative.[63] Thus, Shanks considered the fifteen-minute rule to have "virtually put a stop to news enterprise." It "placed all the papers on a par; the slow ones having an equal chance with the enterprising ones."[64]

The fifteen-minute rule encouraged cooperation on the part of the press. Moses S. Beach attributed the combination of the major New York dailies for procuring the foreign news from Boston Harbor to this rule. Representatives of the leading dailies, situated at the Boston telegraph office, waited night and day for their "fifteen minutes' turn" at the wire. The editors receiving the dispatches, however, soon realized that the reporters "were only duplicating each other's news." The editors therefore decided to "take turns to receive and manifold the news." This arrangement, according to Beach, "merged into the associated press."[65]

Pressures on the leading New York newspapers to cooperate were not limited to the novel circumstances of instantaneous communications. Much closer to home, New York Harbor was long the scene of strenuous competition between New York dailies for scoops in foreign news and shipping reports. Success in this arena required investment in swift vessels and in maintaining efficient crews on constant alert for the arrival of incoming ships. The first evidence of a formal and legal inauguration of an association of New York dailies for acquiring information in common in the age of the telegraph concerned this field of news gathering. In an agreement dated the first of June 1848, five parties—the *Journal of Commerce*, the *Courier and Enquirer*, the *Sun*, the *Herald*, and the *Express*—agreed to combine for the purpose of gathering ship news.[66]

The main feature of the contract was a partnership formed to own and maintain a steamer that had originally been purchased by the *Sun* and the *Journal of Commerce* at an earlier date. To collect recent intelligence from abroad, this steamer, the *Naushon* (renamed *Newsboy*) intercepted incoming ships off Sandy Hook. A small fleet of newsboats, jointly owned and operated by the partners, gathered other shipping news in the harbor. The agreement was limited to news gathering in the port of New York and did not refer to any other mutual arrangements for news gathering on land, at sea, or by airborne birds.[67] This does not imply, however, that no other cooperative arrangements between the contracting newspapers existed at the time. Whereas the harbor association was a legal creature, with a contractual foundation, the contemporary telegraphic arrangements, which involved no property to be held jointly, no questions of maintenance, purchase, or sale of real assets, did not require a binding framework. Thus, an unformulated working arrangement was sufficient.[68]

By purchasing the *Naushon* and establishing the other harbor news arrangements, New York newspapers significantly expanded the scope of the pre-existing association for telegraphic news. The establishment of a better machinery for managing the affairs the newspapers now had in common became necessary.[69] Authority to act on behalf of the association was delegated to committees, on which Frederic Hudson and Henry J. Raymond were active. Moreover, the *Tribune,* which was not represented in the inaugural meeting in the *Sun* office and did not share in the early arrangements of the association for war news from Mexico, became more closely connected to the other five newspapers.[70]

A half year later, early in 1849, the *Naushon* was sold, and the *Tribune* joined the other five members in what remained of their news boat establishment. These changes made it necessary to forge a new and more perfect union of the associated newspapers. A second Harbor News Association was formed on 11 January 1849, with the *Tribune* a full partner.[71]

Thus, in two stages, in mid-1848 and early 1849, the cooperative arrangements between the major New York dailies underwent significant changes. Frederic Hudson, who after the reorganization of May 1848 became actively involved in the expanded scope of news-gathering activities on behalf of the association, dated the origins of the association to this period, 1848–1849.[72] It was convenient for Hudson, a *Herald* man, to dismiss the early stage of the Associated Press, and a faltering memory did not make the events of that period any more vivid.[73] The contractual record at Hudson's hand—the two harbor news agreements, one from mid-1848, the other from early in 1849—probably underpinned his chronology. Yet the pioneering effort of the *New York Sun* and the impression of the first and great "congress of the Republic of news" that gave rise to the New York Associated Press endured. Hudson acknowledged that it took place in the editorial offices of the *Herald*'s great rival, the *New York Sun*.[74]

Finally, neither Beach's memorandum nor the contract of the Harbor News Association—and not even subsequent formal agreements—employed the designation "the New York Associated Press." It was, as noted above, a description rather than a proper name, which was commonly used and eventually caught on. The NYAP, like its name, represented the formalization and institutionalization of de facto practices. It originated as a series of working arrangements, dating from the summer of 1846.

The network of telegraph lines expanded rapidly in the first few years of the Associated Press's operations. New lines brought into the American telegraphic network ever more distant regions. The trunk lines that connected major centers became the focus of vast webs of feeder lines crisscrossing populated sections. More poles carried an increasing number of wires to transact business on the major routes. In the course of the early 1850's the overall transmission capacity of the American telegraph industry increased significantly, as indicated in Table 3. If in the first years of the telegraph, limited capacity prevented competition in telegraphic news gathering, now that telegraphic capacity was dramatically expanded, there was room for renewed competition among daily papers in collecting the latest news. Yet the cooperative arrangements did not give way to renewed attempts at competition.

Clearly, the greater scope of the telegraphic network compounded

Table 3 Increase of Telegraphic Transmission Capacity, 1846–1852

Date	Miles of line	Miles of wire	Maximal yield commercial[a]	Expected yield press[b]
Sept. 1846	1,073	1,815	158,150	47,000
Nov. 1847[c]	3,047	4,700[d]	409,000	122,200
Jan. 1849[e]	6,700	10,200	887,000	265,200
Dec. 1852[f]	16,735	23,281	2,025,000	605,300

a. Using average maximal yield commercial/mile-wire as calculated in Menahem Blondheim, "The News Frontier: Control and Management of American News in the Age of the Telegraph" (Ph.D. diss., Harvard University, 1989), pp. 420–422.

b. Using average expected yield press/mile-wire, ibid.

c. *Source: New York Herald,* 15 November 1847.

d. Estimated.

e. *Source: New York Evening Post,* 13 January 1849.

f. *Source:* Senate ex. doc. 1 (32–2), pp. 571–572.

the difficulty of organizing a comprehensive news-gathering system by single dailies. But perhaps more important by this time was the consideration of price. True, the individual New York daily in the early 1850's could still foot all bills for exclusive telegraphic news transmission. At the time, the entire expense of the Associated Press for telegraph reports was a mere $30,000 per annum—a sum that each newspaper could easily afford.[75] The motivation to share costs may have been predicated on expectations of a dramatic increase in telegraph expenses. Editors foresaw a great expansion in American news supply, in the capacity to transmit it, and in the public demand for telegraphic news.

Newspaper proprietors in the 1840's had only to look at their product to realize that the press was in a state of transition and, moreover, that use of the telegraph was bound to expand. The combining of news received by mail and news received by wire gave the contemporary newspaper a peculiar temporal dissonance. The *New York Herald* of 12 May 1846 reported President Polk's declaration of war on Mexico, "by the Electric Telegraph," on its front page. On page two, a lengthy letter from Washington, written long before the telegraphic communication was received, speculated on the president's next move concerning relations with the neighboring government. Similarly, in the *Herald* of 7 June, a telegraphic dispatch announced the American victory in the battle of Matamoras. Yet in other news items in the same issue there were expressions of concern as to General Taylor's precarious position. An editorial from the *Louisville Journal* of 1847 sarcastically commented on this dissonance: "Having got by telegraph the news of the overwhelming Whig victory throughout New York, we are now daily receiving by mail the papers published in that State just before the election, and we cannot help being infinitely amused at the bold confidence with which the Locofoco papers speak of their party's inevitable triumph!"[76]

The newspaper pages that presented correspondence and exchange paragraphs received through the mails rapidly fell out of step with the expanding volume of telegraphic copy. It is not surprising that the public turned out to be a greater fan of what *Time Magazine* founder Henry Luce termed "fast news," delivered by telegraph, than of "slow news," transmitted by other means.[77] As Horace Greeley testified before a British parliamentary committee in 1851, in America "the quickest news is the one looked to." "The leading article has not then so much influence as it has in England?" inquired the committee. "No: the telegraphic dispatch is the great point," answered Greeley, probably with much regret.[78] Similarly, James D. Reid stated that with the beginning of telegraphic transmission of news, "the public appetite had been whetted, and the chief point of interest was the telegraphic column. There was to be found in the tersest form all the major events by which the world was agitated." Not that the "slow news" of the newspaper was entirely neglected. Once the telegraphic column was read, "the paper was laid aside for the leisure hour of home."[79]

Instantaneous transmission of news, by making knowledge of distant occurrences shared in time, broadened the scope of news that the public found exciting. Because it was in the interest of editors to stimulate the public, they endeavored to attain an ever-increasing quantity of timely news. The outcome of such efforts to include as large a share as possible of "fast news" would of necessity be an all-telegraphic newspaper. As the editor of the *Philadelphia North American* predicted:

> The wires of the Telegraph will be the nerves of the press, vibrating with every impression received at the remotest extremities of the country. The events of yesterday throughout the entire land will be given, as we now give the occurrences at home, to-day. The appetite for news will be whetted into greater keenness by this supply, and those journals which give most, as the *North American* does, will fare best.[80]

The prospect of an all-telegraphic newspaper could have dampened the enthusiasm of even the most affluent editor. With the volume of American news expanding rapidly with the growth of the nation, with more and more localities being brought into telegraphic communication with New York, and with the development of ample facilities for telegraphic transmission, the future promised an exponential rise in telegraph bills. Witnessing the growing public interest in distant events by reason of their timely reporting, the individual editor must have been intimidated by the idea of singlehandedly satisfying an ever-increasing, seemingly insatiable appetite for news. Rising expectations deriving from the telegraph industry and from systematic news gathering thus dictated that cooperative telegraphic arrangements had arrived to stay.

PART TWO

Toward a Monopoly of Telegraphic News

4

"The Advantage of Time": Race for the Foreign News

In the first years after the introduction of the telegraph, the public press as a sector competed with other news suppliers to dominate the flow of information in America.[1] Editors realized that by keeping their power and their resources pooled they stood a chance of extending their news monopoly well into the telegraphic era.

Competition was especially acute when it came to commercial news, particularly that received from Europe. Here, the contest narrowed to attempts to control a thin channel of telegraphic facilities that could be monopolized easily. Control of this strategic junction in the telegraphic network was coveted by many. Owners of the telegraphic conduit between New York and the northeastern seaports, where the foreign news first became available, attempted to control the information that assumed great value by virtue of its rapid transmission over their wires. Telegraphic news vendors also realized that the foreign news, and especially reports of European markets, were an extremely valuable commodity if transmitted exclusively.

The leaders of the New York press soon found that their cooperative arrangement made it possible both to win the race to the foreign news and to retain control of that channel. To an important extent, the leveling of information advantages in America depended on the AP's maintaining a monopoly of knowledge in foreign news.

On the afternoon of 16 November 1846, a petite woman accompanied by a black-eyed girl entered the offices of the Magnetic Telegraph Company in Boston. The woman, dressed in black, handed the clerk a message written in cipher and signed "H. C. Daniels," which she asked to have transmitted to New York. The woman insisted on remaining in the telegraph office until she was assured that her message had reached its destination.[2]

Women did not frequent telegraph offices, and her visit attracted much attention. Because the anteroom was not considered a proper place for a lady, "H. C. Daniels" was ushered, against all the rules and regulations of the office, into the operating room to await the transmission of her message. There she was entertained by all four of the men on duty. Even the president of the line, Francis O. J. Smith, who set the rules, did not seem to mind the woman's presence. Smith had the reputation, later the established notoriety, of exceeding all proper limits and breaking every rule when it came to attractive young women.[3]

The office was anxiously anticipating a rush of messages later in the afternoon, since the steamer *Acadia* had been sighted in the offing of Boston Harbor. One of the swiftest of Samuel Cunard's fleet, the *Acadia* was carrying European newspapers dated less than a week after those brought by previous steamers.

In November 1846, each day of later advices from Europe was crucial. Throughout November, prices of American breadstuffs fluctuated wildly with each new advice from Europe. They shot up on 7 November, when the *Britannia* brought news of the unexpectedly severe crop failures on the Continent and of a commensurate rise in the price of corn. The news brought by the *Great Western*, arriving in New York a week later, dampened the rally in the bull market. From the new advices it appeared that though European shortages were severe, traders in London expected a bounty of imported grains to reach Liverpool, London, and Le Havre shortly. Accordingly, there was not much demand for grain at the high prices that were being asked; buyers preferred to wait and see.[4]

The great cotton market, although generally less fluid and volatile, would also be profoundly affected by the *Acadia*'s news. Cotton prices stood in an inverse relationship to the prices of breadstuffs. The more that Englishmen and Europeans had to pay for bread, the fewer pounds and shillings were available for the purchase of textiles. American farmers and country traders, millers, transporters, and Wall Street magnates alike were anxious to see which way the European markets would finally turn, and they awaited the arrival of the *Acadia* with great suspense. This news was to be similar to the final bell in the seasonal market for grain. It would determine the overall winners and losers of the trade in the crop of 1846.

To the men in the Boston telegraph office, it was apparent that the message signed H. C. Daniels was the harbinger of the new advices from Europe, acquired, somehow, before the ship reached the dock. Written in cipher, the message was obviously confidential. Moreover, it was addressed to Jacob Little, the great Wall Street operator. The woman's insistence that she be assured of the message's receipt in New York indicated that it was of the greatest importance. Her further actions supported these assumptions. When she noticed from her vantage point in the operating room that her message was soon to be received in New York, the woman quickly walked out into the anteroom, and, in obvious haste, scribbled some words on a sheet of paper, which she handed to the girl who had attended her. The girl immediately left the office, billet in hand, and returned a few minutes later in the company of a middle-aged man. After a short conversation with the woman he left the office, and soon thereafter she left too.

About a half hour later, a representative of the NYAP brought a digest of the European news into the office, and the operator began transmitting the dispatch. He had not gotten very far when the telegraph key failed. Communications with New York were down and would not be resumed until the following morning. The line inspectors found that the wire had been severed some four miles out of Boston. Smith and his operators immediately suspected that the woman in the black dress was connected with this sabotage. She appeared to represent the key to solving the mystery of the telegraph interruptions, which occurred with disturbing regularity whenever steamers carrying important European news approached Boston Harbor.

It was not long after the inauguration of the New York telegraph that these disruptions began, delaying what should have been the instantaneous dissemination of European market news. The wires emanating from New York were cut when the *Great Western* arrived in that port early in August; and when the *Caledonia* reached Boston on 18 August, the telegraph was again out of order. "The wires had been cut," surmised the *New York Herald*, "by some rascally speculator" who feared that its use by the public might "upset some speculation in corn or cotton."[5]

The speculator's reason for cutting the wires was quite simple. The press, through its arrangements with Smith, had the privilege of first access to the wires with the arrival of every steamer, as well as the exclusive right to the uninterrupted transmission of the foreign news. If the line was cut, a pre-arranged horse and rail express could speed the news to the conspirators, who in brisk transactions could take advantage of the general ignorance.

Interruptions continued in the following months, preventing first receipt of the news by the New York press. It became common practice for newspapers to announce the expected arrival of a steamer from abroad in the form of a warning to the public. "The *Britannia* is overdue," announced the *Herald* on 5 November, "and the speculators are on tenter hooks. All those having cotton, flour, corn, wheat, etc., etc., to sell, are cautioned against the speculators; and all those wishing to purchase any of these articles must keep a sharp look out to the windward."[6] The newspaper's Boston correspondent informed readers that upon the arrival of the steamer "there are to be all sorts of expresses sent on to New York—horse, foot and dragoons, and he who gets in first is a lucky fellow."[7]

The following day a telegraphic dispatch from Boston asserted that the wires were in perfect working condition. "Arrangements have been made," wrote the Boston reporter, "to watch the flour and cotton speculators to night and if possible, secure their arrest should they attack the telegraph." The editorial staff of the *Herald* in New York concurred. "We shall keep our eye upon the speculators. So will two or three hundred others. We shall then know who cuts the wires."[8] Tensions rose when, on the 7th, no word from the steamer reached New York. The editorials of the *Herald,* by then assuming a moralistic stance, asserted that "speculating in food for the starving millions of Europe and cutting the lightning wires will be considered the very acme of villainy." "The news is now hourly expected," concluded the *Herald.* "Let the public be on their guard."[9] Precautions taken by the telegraph company to prevent another disruption proved to be in vain. The *Britannia* actually arrived early in the morning of 7 November, but because the telegraph wires had been cut, the news was not published by the New York press until the following day.[10] "We understand that a few speculators had the news in Wall Street considerably in advance of the Public generally," observed the *New York Tribune.*[11] When the arrival of the *Acadia* became imminent, warnings were repeated with even greater force in the daily press; but, again, they were of no avail.[12]

The same petite woman reappeared in the telegraph office late in February 1847. This time, too, the wires were cut shortly after she made a series of attempts to transact business with Jacob Little's offices in New York.[13] Needless to say, a steamer was in the offing at the time. This second visit appeared to provide Smith with all the proof he needed of "outrages" committed on his line, and he publicized his findings in the *Commercial Advertiser.* He explained that the stratagem employed by the conspirators was both simple and ingenious. Beating the competition in acquiring the news, they then placed it in the telegraph office ahead of the agent of the NYAP. By cutting the wires after their message went through, they extended the period during which their accomplices were in sole possession of the sensitive information in the unsuspecting markets.

Best of all, Smith's men had succeeded in positively identifying the woman in the black dress. "H. C. Daniels" was found to be Helena

Craig, wife of the Boston news reporter Daniel H. Craig. By linking Craig to the sabotage committed on the telegraph, Smith appeared to have a closed case. Who else besides Craig could beat the enterprising presses of Boston and New York? Who else would have the gall to challenge the NYAP? In the past, Craig had repeatedly and convincingly demonstrated that he had the means to beat all competitors to the foreign news. Moreover, he clearly had the pluck to challenge Francis O. J. Smith's arrangements with as formidable an ally as the powerful press of metropolitan New York.

Like so many other pioneers in the news field, Craig was a New Englander.[14] Born in New Hampshire in or about 1812, he was apprenticed to a newspaper office; but abundant energy and imagination, combined perhaps with a glimmer of genius, drove him from the beaten path of a newspaper career. He early discerned the potential of dealing in information as a commodity unto itself, divorced from the media that published it. Subsequently, when serving the AP, Craig shocked the public by announcing that he viewed news as a salable commodity, no different from "a string of onions." He had no qualms about serving a heterogeneous clientele, ranging from Harnden's express company to the *New York Sun,* and from Wall Street's Jacob Little to the *Herald*'s James Gordon Bennett.[15]

After establishing himself in Baltimore in the late 1830's, Craig experimented in training African pigeons to deliver information. He employed his birds in carrying Washington news to the *Sun,* an enterprising and popular Baltimore penny newspaper.[16] Later he shifted his operations to Boston and, years before the establishment of the telegraph, made a specialty of anticipating the European news brought by the Cunard fleet. Craig boarded the Boston-bound steamers at Halifax, their first port of call in the New World. During the early part of the two-day voyage to Boston, he leisurely digested the news from Willmer and Smith's authoritative *European Times* and other European newspapers. He then printed the news, in fine type, on slips of tissue paper. When the shores of Massachusetts came into sight Craig would dispatch the pigeons laden with news to his Boston accomplices, who forwarded it to the final customers, several hours before the ship tied up at its Boston wharf.

The New York press complained about this practice to the management of the Cunard line. On a subsequent voyage the captain confiscated Craig's cargo of African pigeons. "Apprehending this outrage," Craig had tucked one of his birds into the pocket of his overcoat before boarding the ship. After printing the news and fastening it to the bird's leg, Craig later related, "I went on the deck and flew the bird close to the captain's head. He darted into his stateroom and caught his rifle, but before he got a chance to shoot, the bird was a mile above him, flying straight to his home in Boston, a hundred miles away." The winged envoy delivered his news "several hours in advance of the steamer."[17]

Thus, in both ability and temperament, Craig fit the role of culprit in the interruptions of the telegraph wire. Craig himself proudly admitted he had beaten the New York press to the *Acadia* news. Yet he flatly denied responsibility for cutting the wires. His version of the story carried a strong implication that Smith and his telegraph line were performing some functions additional to those of a common carrier of information.

In a letter to N. B. Carney, Jacob Little's clerk, Craig confirmed most of the details of Smith's story, with one important exception: the message to Little was not transmitted to New York, even though the line was in working condition. Helena Craig, according to this version, reached the telegraph office with the foreign news two hours before the steamer was even signaled as approaching the harbor. Her instructions were to send the news to Jacob Little, and if for any reason the news could not go through to New York before the steamer docked, she was to deliver it to a local customer. At the telegraph office she was put on hold for hours, "waiting for the lieing [*sic*] knaves to *'get a current.'*" When she "suspected from the movement in the office" that the *Acadia* was entering the harbor, she quickly wrote out the leading items of the news and sent them, "with our little girl," to the Boston customer.

The latter came charging to the telegraph office, "as he could not make out her hasty *pot-hooks.*" This, concluded Craig, "is the head and front of all the direful plottings and counter-plottings of these two parties against the general peace and order of all the good people of Gotham and Yankeedom and of Francis Orville Jonathan Smith in particular."[18]

For years Smith repeated the charge that Craig was guilty of foul play. When he made it before a committee of the Massachusetts Senate in 1859, Craig sued Smith for libel. At that time the committee was considering a bill of crucial importance to the telegraph interests with which Craig and the NYAP were affiliated, and the charge was reiterated, over Smith's name, in the local press. At that juncture, Craig could ill afford to let the charges go unanswered. It took the Superior Court of New York almost three years to sort out the affair, and after hearing thirty-four witnesses for the plaintiff alone, besides taking dozens of depositions, it found in favor of Craig.[19]

The New York press reached a similar conclusion in much less time. Notwithstanding past clashes with Craig, they did not believe him responsible for the crime. The press had their own hunch as to who was responsible for the telegraph interruptions: their prime suspect was Fog Smith himself, manager, superintendent, and majority owner of the telegraph line. They believed that Smith was in the business of selling telegraphic reports of steamer news to New York speculators and that he declined to send competing reports under the pretext that the wires were down, so as to ensure exclusive information for his customers. Smith manufactured the case against Craig, the New York editors believed, in order to divert suspicion from himself. Even Smith's wordy statement about evidence he uncovered in the case did not convince the *New York Sun* of his innocence.

> It is almost unnecessary to say that we don't believe a word of it. We know so much of that Boston telegraph, the stories manufactured to order, and the outrageous manner in which it is conducted, that we are prepared for any "statement" that will shield the line from deserved reprobation. "Unknown persons" cutting the wires, indeed! The *infamous management* of *that line* stamps this latest fabrication with the mark of a well known authorship.[20]

The *Sun* was alluding to Smith's authorship of an earlier statement relating to the irregularities in transmitting the *Britannia*'s news on 7 November 1846. On that occasion the line was severed before the press dispatch of the foreign news went through. The *Sun*, by a series of sworn affidavits, established a solid case against Smith's line for suppressing the message and giving undue priority to a private speculator. The *Sun* presented an air-tight case that demonstrated either "gross mismanagement" or, as the editor appeared to believe was more likely, "foul play."[21]

The *Sun* called for an investigation and coupled the demand with an unconcealed threat: "We do ask fair play and we will have it. The abuses of this great vehicle of public intelligence must and shall be corrected or the companies entrusted with its management shall cease to exist. Our state legislators which granted them the corporate privileges can punish their faithlessness, as well as protect the wires."[22]

The press was not satisfied with Smith's reply to the accusations, published in the *Journal of Commerce*. In these remarks, perceived by the *Sun* as an "attempt to mistify and be*Fog* the vision of . . . readers," Smith, obviously indignant, denied any malpractice by himself or his subordinates. "His wrath boils over so much," observed the *Sun*, and "his remarks run so wild [that] we think it best to let his observations go for what they are worth," which it obviously did not consider very much.[23]

The *Herald* noted that public confidence had been shaken. "The directors of all companies," it predicted, "will be obliged to use every expedient that can be devised, to prevent not only breaches of the wire, but also breaches in confidence, of the latter of which we have recently had some instances."[24] Alluding again to Smith's line, the newspaper argued that "when such an important method of communication as this is in the hands of those who sacrifice public benefit for private interest, it becomes a dangerous matter."[25] The *Herald* too issued a warning to Smith: "The conductors or proprietors of these telegraph lines no doubt imagine that they are beyond the reach of the people or the power of the press, but they may find out their mistake when it is too late."[26]

Viewed from Philadelphia, the battle lines in the struggle to control the foreign news were somewhat hazy. From that distance, the chief contenders appeared to have much in common. "A war has broken out between some of [the New York morning papers] and the management of the Boston line of Electric Telegraph," noted the *Philadelphia Public Ledger.* Although the newspaper considered the underlying issue in the war—the public control of news—to have much merit, it did not find that merit in either party to the controversy. The contest between F. O. J. Smith and the NYAP was "merely a war between two sets of monopolizing speculators." Smith appeared to the *Ledger* to have the upper hand at the time. The paper related that "the now grumbling party," namely, the five or six large dailies composing the association, "first obtained a monopoly of the telegraph, but have been crowded out by the other party who now hold possession."[27] But, as the *Public Ledger* came to learn in a quarrel with the same speculating party a decade later, the NYAP was not an easy institution to crowd out.[28]

The forces that eventually dominated the struggle to control telegraphic news flow were remarkably well represented in the minor if lively conflagration of November 1846. The same fundamental issues were to recur, in various guises, in the first two decades of American telegraphy. Control of the news from Europe remained the focus of the telegraphic news drama for years. Even the major performers in the incident continued to occupy center stage in subsequent acts.

Each of the three major contenders in the conflict of November 1846 represented a distinct power in the general contention for a monopoly in time over the news. Daniel H. Craig found ways of getting to its sources first; Fog Smith had a monopoly on the most rapid means of transmitting it; and the NYAP played a leading role in its dissemination.

Smith entered the struggle with a monopoly in its strictest sense: "a firm that is the sole seller of a product for which there is no close substitution." As in other cases of pure monopoly, so rare in American history, he had a special advantage conferred upon him by law—that of an exclusive patent.[29] He held the sole rights to use Morse's patent in the Northeast and had full charge of the system that employed the invention, as well as the capital invested in it.

The NYAP commanded the combined resources of six institutions of great wealth and of considerable social power and political leverage. It claimed to represent the people in the contest for control of news. Besides defending the public's interests, it was perceived to have a strong influence in shaping public opinion. The AP was a dominant factor in the consumer market for news. In contrast, Daniel H. Craig had a petite wife, a little girl, and a few useful personal qualities. He demonstrated foresight, great organizational abilities, a fertile imagination, and remarkable perseverance. Born and bred a New England Protestant, he had a propensity for hard work and apparently boundless resources of energy. As it turned out, these last two bases of power proved the more effective. Yet Smith's position was formidable enough. He did not give up his potential monopoly of the foreign news without a fight, during the course of which he gave Craig and the AP all they could handle. First, however, the NYAP squared off with Daniel H. Craig.

Craig, through the use of his carrier pigeons, often beat the NYAP to the telegraph office with the foreign news.[30] He supplied the news to speculators, but he soon found that after market operators reaped their full advantage, he could still sell the news, second hand, to the yet ignorant press. New York newspapers that were not members of the AP joined together to receive Craig's news collectively.[31] Since a share of the income and much of the prestige of the AP newspapers depended on their being the prime source for the latest news, the competing association represented a serious threat. Provided with the earliest news, it stood a chance of superseding the AP and seriously damaging its member newspapers.

Fortunately for the latter, the opposing association did not succeed. Craig had no particular interest in sustaining a rival organization, especially one made up of the less affluent presses of New York. Since his independence allowed him to charge each newspaper separately for the news, he had a definite interest in the welfare of the dailies that constituted the AP. These newspapers could pay him the price he considered proper recompense for his labors to beat their cooperative effort.[32] In cases when his service beat their own, Craig sold his news to the NYAP newspapers "at a fabulous price," and to the outside press as well.[33]

The news the AP got in this manner was not the first news, was not exclusive news, and was expensive. After paying Craig for his services, the NYAP was still compelled to pay Smith for a 3,000-word dispatch upon the arrival of every steamer, no matter how stale the news was by the time it reached New York over his line, which was so subject to interruptions.[34]

The AP newspapers resorted to expensive expresses between Boston and New York when repeated experiences proved that their privilege in using the telegraph amounted to little when the wires were cut. But expresses alone were not enough. The NYAP soon realized it would have to beat Craig at his own game and capture the news at its source.

Their first significant effort to get the news to the wire-head before Craig proved a disaster. Late in April 1848, the association chartered the steamer *Buena Vista* to run the news from Halifax to Boston. The Cunard steamers usually took almost a whole day at Halifax for debarkation, embarkation, and refueling. That delay could be avoided by dispatching the news to Boston on board an express steamer immediately after receipt of the foreign newspapers.

The steamer the association chartered, at about $2,000 for each run, itself represented problems. "Puffed in the New York papers as the greatest clipper of a steamer that ever walked the water," noted an amused Halifax newspaper, the *Buena Vista* set out from Halifax for her maiden voyage on 11 May 1848 carrying a budget of European news brought over by the *Cambria.* "Away sped the *Buena Vista* down the harbor as fast as she could scud," just to return a few hours later "hobbling up the harbor like a lame duck." Only after two false starts did the steamer finally get under way to Boston. "In place of announcing the arrival of the *Cambria,*" scoffed the provincial newspaper, the anxious New York editors "will be sadly at a loss for some time to know what had happened to prevent her arrival at Boston two or three days after the *Cambria* reaches New York." The *Buena Vista*'s second voyage from Halifax, on 24 May, was also beset with mechanical failures. The Cunarder easily beat the express steamer, despite the latter's considerable head start.[35]

A subsequent, even though more speedy, voyage disappointed the New York editors. For not only their choice of the ship but also the very idea of employing it was flawed. Indeed, Craig was delighted with the New Yorkers' hiring the *Buena Vista.* One of the few who had confidence in the safety of passage on the unlucky express steamer, he boarded it at Halifax with his pile of European newspapers and a basket of birds.[36] The steamer saved a day or so for both competitors, but again the birds beat the NYAP agent to the Boston telegraph office by the usual margin.[37] Yet there were ominous signs for Craig in the hiring of the *Buena Vista.* Not only did it demonstrate the immense resources of the NYAP, but, even more threatening, it signaled the association's determination to control the foreign news field at any cost.

The same power and determination soon exhibited themselves in a much better conceived plan that seriously threatened Craig's position. In 1846, the British North American Maritime Provinces began considering propositions for the establishment of telegraph lines through their territories. Interested parties in Quebec and Montreal (later represented by the British North American Electric Telegraph Company) lobbied strenuously for an east-west orientation for the line, so as to connect Halifax with Montreal.[38] Yet the provincial governments of New Brunswick and Nova Scotia both knew that their interests would be better served by a line connecting their provinces with the great mercantile centers of the United States.[39] Nova Scotia could give lip service to the cause of "protecting this portion of the Empire" and to efforts "to unite the Colonies more closely together" by feigning support for a western connection; either way the line from Halifax would follow the same route.[40] The decision lay squarely with the Province of New Brunswick. The New Brunswick House of Assembly on 4 October 1848 granted a charter to the company prospecting a line over the southern route, which would connect with the American seaboard lines.

The choice of the southern route was based on the tacit understanding that the NYAP would employ the provincial lines for transmitting 3,000 words to New York upon the arrival of each steamer.[41] Cursory cost accounting demonstrated that if the AP paid the same pro rata tolls it was charged on Smith's lines from Portland to New York, this would defray virtually the entire expense of operating the provincial lines. The projected telegraph companies were thus assured of breaking even before sending a single commercial message. Work on the New Brunswick line began without delay.

Yet there were still thousands of poles to erect and hundreds of miles of wire to stretch before the telegraph would reach Halifax. With each advance of the wire, Craig pushed his base of operations further north.[42] On 13 February 1849, the line of the Maine Telegraph Company connected Calais with Portland, the terminus of Smith's line. The next segment, from Calais to St. John, New Brunswick, had actually been completed a little earlier, on 1 January 1849.

With St. John directly connected by telegraph to New York, Craig (in Boston at the time) arranged for running what became known as the "Halifax Express." The news received at Halifax was to be forwarded by horse express to Digby, then, by a waiting steamer, across the Bay of Fundy to St. John, where it would be telegraphed to New York. With this imaginative, if extremely costly, plan, Craig was sure to maintain an uncontested monopoly of the European news.

Word of Craig's plan leaked to the NYAP.[43] The association, determined to foil him at any cost, dispatched its Boston agent, John Turel Smith, to the British Provinces. He set off immediately, hoping to arrange for the best horses and fastest steamer before Craig did. It was never easy, however, to outspeed Craig, who overtook Smith at Augusta and passed him at Bangor. Arriving first at St. John, Craig succeeded in chartering the fastest steamer available for crossing the Bay of Fundy.[44]

At this point, however, a new obstacle was thrown in Craig's path. With the completion of the St. John-Calais segment, the New Brunswick Electric Telegraph Company had exhausted its resources. A plea to the Provincial Government for aid in pushing the line further north, to the border of Nova Scotia, was rejected.[45] Lawrence R. Darrow, the manager of the line, who invested heavily in it, appealed to the NYAP for support. With Craig in mind, it readily guaranteed payment of double the premium commercial rates for 3,000 words, upon the arrival of every steamer, whether it sent a dispatch or not. The AP's only demand was priority over all other customers for uninterrupted transmission of the foreign news. The desperate Darrow convinced the managers of the Maine line to accede to the association's stipulation, and he received a promise from the Nova Scotia authorities to concur, once their line was completed. On 9 February, an agreement to that effect was signed, and on the merits of the AP's guarantee, the subscription books of the New Brunswick company were rapidly filled.[46]

When Craig entered the telegraph office in St. John to confirm his arrangements for the Halifax Express, Darrow, now bound by a contract with the NYAP, recommended he save his money and abandon the express. Under no circumstances would he permit Craig to telegraph the news "until after the agent of several of the New York journals had sent the news forward." "I answered him," Craig later testified, "that I expected to beat all competitors in putting the news on file; that if I should do so, and he left me in any doubt as to my having the first use of the wire, I would prevent him from sending the news to any one, and then prosecute him for damages." Most likely it was the first of these two threats that was more persuasive. After all, the controversies with F. O. J. Smith had made Craig's supposed ability to prevent others from using the wires widely known.[47]

On 21 February 1849, the *Europa*, eleven days from Liverpool, arrived at Halifax, and Craig ran the Halifax Express for the first time, with no serious opposition. The horse express covered the distance of 144 miles from Halifax to Digby in about eleven and a half hours. But on reaching Victoria Beach, Digby Gut, Craig later related, "I found my express steamer forced down the Annapolis river and firmly imbedded in the ice, five miles in circumference and eighteen inches in thickness." It was only after "many hours of hard labor in cutting an opening for the steamer" that the express could proceed.[48]

The steamer finally arrived safely at St. John, and at the door of the telegraph office Craig was given a cordial reception by the superintendent, as might have been expected. The news was telegraphed to New York only twenty-four hours after the *Europa* reached Halifax. Hard work had its rewards, and, once the news was received in Boston and

New York, Craig enjoyed the satisfaction of having his agent sell it to the NYAP. "It is quite certain," offered Craig, "they paid more for it than they ever paid before or since for any two columns of news by telegraph."[49]

By the time the next steamer was due at Halifax, John T. Smith had perfected his own arrangements on behalf of the NYAP. Halifax held its breath when on 8 March the *America* neared its wharf. With its purser unloading two packages of the freshest European newspapers in North America, the race between the New York press and Daniel H. Craig began. The first leg, from Halifax to Digby, was won by the AP. Its horse relays reached the Bay after eight hours, twenty-seven and a half minutes of galloping. Craig's rider pulled in two and a half minutes later. But his dispatch was immediately placed aboard the *Conqueror*, its engines already roaring, while the AP rider looked on, dispatch in hand. His steamer, the *Commodore*, was nowhere in sight. The second lap of the race belonged to Craig.[50]

Craig's Halifax Express continued to run whenever a Cunarder reached Halifax, until Halifax and St. John were connected by telegraph on 9 November 1849, but it ran on behalf of the NYAP. The demonstration in the Bay of Fundy was enough to convince Henry J. Raymond and Frederic Hudson, the executive committee of the NYAP, that Craig was too dangerous an adversary to roam the Maritime Provinces representing the interests of Jacob Little or a competing association of metropolitan dailies. As Lawrence R. Darrow was to remark, "Letting such a man as Mr. Craig, whose energy and activity we cannot destroy, or cripple, be let loose . . . looks to us like utter madness."[51] Craig was summoned to New York and asked to become the agent of the NYAP in charge of the foreign news.[52]

By this time Craig—like his European counterpart, Julius Reuter—recognized that there was no future for the carrier pigeons that had been the original basis of his success. With the gap between the steamers and the telegraph office narrowing daily, Craig realized that control of the wire-head was becoming the prime factor in the race for the foreign news. The AP held that power, and he accepted their offer on the spot.

Craig, as agent of the NYAP, had full charge of the association's operations in the foreign news field. It was up to him to decide on the means of procuring the news, and it was his prerogative to make the arrangements for transmitting it. Francis O. J. Smith had to iron out with Craig the differences between his telegraph monopoly and the powerful press of New York. This combination promised trouble, and it was not long in coming.

Members of the NYAP had not entirely forgotten Smith in their preoccupation with their contest with Craig. On the contrary, he was receiving continual attention in their newspaper columns. Although they initially concentrated on challenging Smith's management of the Boston line and expressing their continued suspicions that he was benefiting from private speculation in information, they soon found a grander cause. It was provided courtesy of Henry O'Rielly, the indefatigable promoter of the telegraph in the West.

Soon after O'Rielly began developing his western telegraph empire, he fell out with the Morse patentees. In one isolated display of harmony, Morse, Kendall, and Vail were joined by Smith in common antagonism toward O'Rielly and in efforts to defeat his telegraph interests.[53] Kendall hoped to destroy O'Rielly by granting patent rights to entrepreneurs—the pioneer telegrapher Ezra Cornell among them—who were jealous of O'Rielly's initial success and tried to establish lines in competition with his system. Kendall knew that competition between rival Morse lines over the same routes was a prescription for failure. He realized that it would bring chaos into a system that more than anything else needed coordination and unity. Yet he considered a flourishing O'Rielly telegraph empire a greater evil.

Fog Smith was among the first to take the field against O'Rielly in the Northwest. The telegraph wars in the Mississippi Valley and the Great Lakes region indeed brought the country's fledgling telegraphic system into disarray, but they did not subdue O'Rielly. Rather, they exhausted all those who entered the competition, including Smith. Viewing the destruction of O'Rielly as the end that justified all means, Smith forbade communications to or from O'Rielly's western system to pass over his seaboard lines. Smith's policy hurt O'Rielly, whose lines depended heavily on through communications with the urban centers of the Northeast. O'Rielly responded by abandoning his efforts to reach a negotiated settlement and instead embarked on an all-out effort to defeat the Morse patent. A Jacksonian of the old school, he presented his cause as a struggle against monopoly and for equal rights in telegraphy. "Monopoly" proved a splendid battle cry.[54]

O'Rielly unleashed his well-lubricated propaganda machine as the vanguard of his effort. Articles and letters were submitted to editors that then, in the form of printed extracts from newspapers, appeared in numerous pamphlets he issued. Completing a recycling process, the newspaper articles appended to the pamphlets were widely quoted by the press.[55] As it turned out, O'Rielly's years of pursuing a liberal policy toward the press paid off in bountiful dividends. The press issued a chorus of indignant cries against Morse's monopoly and heaped abuse on his chieftains, Kendall and Smith.[56]

The metropolitan press of New York took up, with great enthusiasm, the cause of defending the public against the scheming monopolists of timely information. "The monopoly of intelligence," exclaimed the New York *Express* in perhaps the most high-minded specimen of the genre,

> is too great an element of power to be vested in a human being. The magnetic telegraph is the application of electricity, with lightning speed, to the transmission of thought . . . To intrust such a power to a person, or persons, is to intrust them with the highest exercise of authority over the human race . . . Here, of a sudden, has crept in upon us an uncontrolled and illimitable Power, before which all others fade into insignificance and contempt. To allow such a power to grow up, and to strengthen itself, in our apathy, is to commit it to the monopoly of light, of news, of intelligence, of speculation, of trade, of commerce, nay, of the highest and dearest interests of the human race.[57]

Journalists went beyond abstract reflections on monopoly over "this great instrument of *Thought* and *News*." They provided some illustrations of the telegraph's potential for evil "in the hands of bad men" or men "who consider it *their* property, with which they have a right to do as they please." These concrete examples were generally based on their experiences with Fog Smith and his high-handed management of the Boston line.

Sidney Morse, the inventor's brother, pointed out in his *New York Observer* that the press was not altogether a disinterested party in its campaign to rescue the public from his brother's "odious monopoly." Competition in telegraphy would enable newspapers to receive their telegraphic news much more cheaply. His fellow editors responded, however, that they found nothing inconsistent with the public good in that.[58] Yet they did, in fact, have an ulterior motive for defeating Morse and particularly Smith: they were striving for a monopoly of the news themselves.[59]

With the aid of the metropolitan press, O'Rielly convinced the New York State legislature to pass, on 12 April 1848, what amounted to a general law for the incorporation of telegraph companies. O'Rielly was greatly assisted in this effort by James Brooks, editor of the New York *Express*, an AP newspaper. Brooks's legislative fingerprints were noticeable in a unique clause of this act. The clause required telegraph companies to transmit messages in the order in which they were received; but allowed an exception in the case of the public press "for the transmission, for the purpose of publication of intelligence of general and public interest, out of its regular order."[60]

The most decisive move in O'Rielly's war on Morse's monopoly was an attempt to introduce alternative systems of telegraphy.[61] Hugh Downing, one of O'Rielly's lieutenants in the West, secured the rights to the printing telegraph invented by Royal E. House. Downing received great encouragement from the press when he established a company to build a line employing House's printing telegraph between New York and Boston.[62]

O'Rielly himself took to the field with a much more dangerous weapon. It was the Bain electro-chemical telegraph (forerunner of the present-day facsimile). In experiments it demonstrated outstanding potential for rapid transmission of long messages, and that was precisely what the press required.[63] To an uproarious press, O'Rielly announced his plans to construct a Bain line between New York and Boston in competition with Smith. The route thus became "the telegraphic race-course of the country."[64]

The enthusiastic endorsements of O'Rielly's plan by the AP newspapers were not altogether spontaneous. The association had in fact asked O'Rielly to build such a line. In the course of negotiations, the AP unanimously declined a proposition to invest directly in the line but decided, again unanimously, to guarantee O'Rielly all the business of the AP on that route for at least six months, should O'Rielly build the line. He was given this resolution in writing, as an inducement to prospective investors.[65]

Smith realized that with a competing line, not only the luxury of news control, but the actual existence of his telegraph company was at stake. If the press business passed over to O'Rielly, it was questionable whether private business alone could cover his line's operating expenses. Smith apparently did not know of the AP's commitment to transact its business over the new line. Yet he knew very well that with Craig making the decisions regarding the AP's eastern operations, his chances of retaining this pivotal source of income were almost nil.[66]

Smith met this emergency with a far-reaching plan. He would put the leverage of his 350 miles of yet unopposed telegraph lines behind a competitor of Craig's. At best, this loyal agent, aided by "superior facilities" on Smith's lines, would supplant Craig as the AP's agent. This would ensure Smith's keeping the association's business. In the event that it did not do so, the agent, because of Smith's control of the wires, could supply a better foreign news service that would ultimately supplant the service of the NYAP. Once he had his own wire service providing the European news to the national press, Smith would fear no competition to his lines. Both O'Rielly and Downing were in dire financial straits; they could not possibly sustain their competition for long without the patronage of the press.

Smith decided to make his move once the wires reached the harbor of Halifax. Shortly after 9 November 1849, when Halifax was brought into direct telegraphic communication with New York, John T. Smith, the former Boston agent of the NYAP, appeared in Halifax and rented an office near the harbor.[67] In a card published in the Boston papers, Fog Smith informed the public and the press that he "had encouraged" his namesake to establish a "reliable commercial agency . . . to act conjointly with the Telegraph." With what proved to be considerable foresight, Fog Smith conceded that the undertaking was "by no means an easy one," yet he granted the agency "exclusive privileges" on his lines. He assured the public that "every advantage which the Telegraph Lines under my direction can afford, will be at all times most freely awarded" to John T. Smith.[68]

During four years of constant observation, the two Smiths had learned a thing or two from Craig. He taught them the significance of first possession of the news and the importance of rapidity in feeding it to the telegraph line. Craig, perhaps, underestimated the learning ability of his adversaries.

The newly established Nova Scotia telegraph, the only government-operated telegraph line in North America, adhered to the fifteen-minute rule and to the principle of "first come first served," with one important exception. The AP had the privilege of an uninterrupted transmission of three thousand words upon the arrival of each European steamer, from the moment its dispatch was submitted.[69] The agreement was moot as to the rights of competitors who reached the office after the arrival of the steamer but ahead of the NYAP. Would their communication be transmitted? Moreover, there was no specification of who the authorized agent of the NYAP was. It was primarily through the latter ambiguity that John T. Smith planned to foist his reports on the NYAP.[70]

When the first European steamer to reach Halifax since the appearance of John T. Smith neared its wharf, two of the fastest teams of horses in Halifax were waiting on the dockside, together with half the local inhabitants as spectators at what promised to be the best show in town. One team of horses served Craig, the other was to run for John T. Smith. As Daniel H. Craig was absorbed in watching the final maneuvers of the pilot, waiting for the opportunity to board and seize his package of European newspapers, the purser descended to the steamer's paddle box and tossed a packet of newspapers into the waiting hands of Smith. By the time Craig boarded the steamer and received his package of newspapers, John T. Smith was inaugurating his foreign news service over the wires of the Nova Scotia telegraph.[71]

John T. Smith's first dispatch was addressed to the NYAP, and for that reason it was given the privilege of uninterrupted transmission. Craig protested to the Nova Scotia authorities, claiming that he was the association's authorized agent and only his messages were entitled to priority. The New York committee agreed with Craig and informed the Nova Scotia telegraph commissioners that in the future they would not accept (or pay for) messages sent by John Smith.[72]

Fog Smith realized that his agent would require a more forceful introduction to the NYAP and resorted to his ambitious fallback plan. It called for the formation of another newspaper association to compete with the AP. John T. Smith, who now was in control of the first European news, would be its agent. The competition would either force the NYAP to terms or replace it as the leader in foreign news. The first step in this sweeping plan was to create dissension between the press associations of Boston and New York. Smith charged that the New Yorkers released the foreign news in New York City before the Boston press was allowed to publish it, thus giving their local traders, brokers, and merchants, let alone journalists, a tremendous advantage. This was a potentially popular cause. Bostonians were notoriously jealous and resentful of New York, ever since Gotham supplanted the Hub as the commercial capital of the country. Smith, however, was only partly successful. Most of the Boston editors did not believe the news was first released in New York and remained loyal to the NYAP.[73]

But Smith did succeed in driving a wedge between the morning and evening newspapers of Boston. The telegraph, by hastening news and making it available at any hour of the day, encouraged morning newspapers to publish afternoon editions. The exclusively evening newspapers were badly hurt by this development, and Smith took advantage of the rift within the Boston press. With the aid of the malcontent Charles O. Rogers, editor of the *Boston Evening Journal,* he succeeded in breaking up the Boston association, on 27 November 1849, and forming a separate association for the evening papers.

It was with lingering doubts as to his ability to get to the news first that John T. Smith, now agent of the Boston *Journal, Transcript,* and *Traveller,* awaited the arrival of the next European steamer early in December.[74] Craig, who was responsible for those doubts, was there too, representing the morning press of Boston and the AP of New York. Although Craig was apparently no more confident of his ability to win the race to the news, he probably succeeded in hiding his anxiety. Craig, observed one of his contemporaries, "preserved at all times the placidity of a summer's morning," yet "no man with a face so calm and a manner so suave ever stirred a pot so lively."[75] It was this composure that enabled Craig to play the bluff so well.

Soon after his defeat, Craig began circulating rumors that upon the arrival of the next steamer, arrangements he had supposedly made some weeks earlier would assure him of getting the news ahead of Smith. On the eve of the steamer's expected arrival, Craig positioned a phantom newsboat five miles down the harbor, with a fast horse express at a beach nearby. Another newsboat was located a mile from the steamer's wharf, opposite the telegraph station. When Craig assembled a dummy package of easily recognizable European newspapers with their dates carefully concealed and then handed it to a reliable accomplice, the stage for the sham was set.

Both the psychological warfare and the instruments of deceit proved fully successful. As the Cunarder steamed up the harbor, Craig's agent on the waterfront momentarily immersed the dummy package of newspapers in the waters of the Atlantic, then ran with the dripping bundle to the telegraph office. "Here is Craig's parcel of European news," he announced, placing the wet package on the office table. A quarter of an hour later, John T. Smith, after a quick gallop from ship side to the telegraph office, laid his own package of newspapers on the desk. The clerk, pointing to the wet package, informed Smith that Craig had beaten him again. "With a single glance at the dripping package, and a fearful malediction upon his competitor, Smith turned upon his heel and drove to his hotel." He was last seen in the vicinity of Halifax boarding the same mail steamer on the final leg of its voyage to Boston Harbor.[76]

With Craig's victory, an important element of Fog Smith's plan had miscarried. His pet newspaper association lost its main advantage, namely, early receipt of the news. Moreover, the NYAP no longer had any reason to replace Craig as its agent. To keep the foreign news flowing through his own wires, rather than O'Rielly's, Fog Smith would have to overcome the NYAP in an all-out war. And he went right ahead with it, immediately abrogating his contract with the NYAP, the breakaway of the Boston evening newspapers serving as a pretext. Smith then called on the managers of the other lines between Boston and Halifax to follow his example and nullify their contracts with the NYAP.[77]

The manager of the New Brunswick company responded to Smith's call by acknowledging his debt to and dependence on the NYAP. "Without the sum paid by the press," he wrote Smith, "it would be impossible for us to support ourselves under the best of circumstances."[78] But the Nova Scotia telegraph, established for the benefit of the people, its management scrutinized by a watchful House of Assembly, had to answer to higher principle. The priority the Nova Scotia authorities granted the AP was based on the premise that the association represented the public interest. It was the people, through their press, who received priority over the few, selfish speculators. Accordingly, the

AP's privileges were contingent on two important conditions: the newspapers receiving the news had to place the important price quotations on their bulletin boards immediately upon receipt; and any newspaper that paid its fair share had to be allowed access to the early news. This was a heavy concession forced upon the NYAP to assure its ascendancy. The association was even compelled to extend its privileges to two additional New York newspapers.

Smith now claimed that the NYAP had not complied with this requirement, and he presented the association of Boston evening newspapers to the Halifax commissioners as a legitimate competitor to the NYAP and its allies in the morning press of Boston. Although a small organization, the Boston association had the advantage of employing an honest agent, John T. Smith, and had nothing to do with Daniel H. Craig. It was with great energy that Smith went on to make this latter point, in an all-out effort to discredit Craig.

The campaign to discredit Craig at first appeared easier than expected. Late in November, the authorities of New Brunswick detected and arrested a man named John Anderson in the act of cutting the telegraph wires. A small-time utility man, Anderson clearly was working in the interests of other, more powerful parties. As Anderson's employer, Craig was an ideal suspect. Craig was the one man who had the steamer news ahead of the NYAP. Betraying his employers, he could withhold the press dispatch until after he sent the news to speculators. Cutting the wires after the news went through was, of course, a natural adjunct to such a scheme. As if to compound Craig's difficulty, it was found that he had employed Anderson in the past for delivering news.[79]

Were that not enough, Craig, obviously shaken by John T. Smith's short-lived victory, decided to deter his rival from further efforts by displaying his ultimate weapon—the carrier pigeons. Craig imported his loyal birds to Halifax and found the roof of the telegraph office the best place for them to roost. This popular species of bird, during the course of its long history in the service of man from the days of the Romans to the battle of Waterloo, had become associated with speculation. The combination of Craig and Anderson, carrier pigeons and telegraph office, was too suspect to go unnoticed.

To make matters even worse, Craig wrote John T. Smith a threatening letter, promising to beat him by any means, fair or foul, should he return to Halifax endowed with the powers that his master, Fog Smith, conferred on him. Fog Smith had this private letter published in the Boston evening papers. Craig's lively vocabulary and dark insinuations made the letter "a fruitful subject of comment, from Halifax to New Orleans."[80] Yet the smear campaign against Craig did not succeed. Horace Greeley, James Brooks, James Gordon Bennett, and the other members of the NYAP stood squarely behind him. Lawrence R. Darrow, superintendent of the New Brunswick line, after a careful investigation, exonerated Craig from any connection with Anderson in cutting the wires.

Craig's own defense was brilliant. He began by pointing to the probable culprit. Late in 1847, Ezra Cornell established a system of visual telegraphs between Halifax and Boston in the interest of a Wall Street operator named Norton.[81] Faster even than carrier pigeons, it was apparently used repeatedly, weather permitting. Once the electric telegraph reached Halifax, the operators of the optic telegraph could succeed only if they cut down the wires of the electric telegraph.

Craig also appealed directly to the public, in a moving card published in the Boston press. He presented himself humbly (and accurately) as the little guy, who had entered the news business trying to make a living for himself. Because of his early success he incurred the wrath of giant, privileged powers. These powers were now trying to destroy him. Craig vehemently denied the accusations against him. They were "maliciously, basely, and utterly false—false as the heart of falsehood could frame it—'a lie! a d—d lie!' "[82]

His best defense, however, was his pointing to Smith as the man responsible for the "meanness, falsehood, and knavery" of those accusations. Smith's public image had become so tarnished as to make his righteous indignation toward speculators somewhat suspicious. It was no surprise that the public took Craig's word over Smith's. It may also have been that the public appreciated Craig's courage. If Smith and Craig were perceived as two rascals squaring off, Smith was the villain the public loved to hate, and Craig was the one they loved.

But behind the ultimate failure of the campaign to discredit Craig lay the power of the New York press in the arena of public opinion. Smith's appeal to public sentiment was ineffectual from the start. There was no way he could persuade the newspaper-reading public over the heads of the editors of their newspapers. And it was the press that helped make Smith's reputation so notorious. The editors of the New York press were, therefore, not at all concerned about the public's possible disapproval of their employing Craig.

Smith's legalistic sophistry on the rights of the Boston evening papers was no more successful. As he should have known, his case was much too thin. The lieutenant governor of Nova Scotia and its House of Assembly confirmed the decision of the telegraph commissioners to continue the preferential arrangements with the NYAP. They would grant no special offices to the association of the Boston evening papers or to its agent, John T. Smith.[83]

It was only after his decisive failure to secure the press business by stratagem that Smith decided to combat O'Rielly's competing line in court. On 19 December 1849 he filed suit against O'Rielly for use of the Bain telegraph instrument, claiming it was an infringement on Morse's patent. Smith knew that the suit could drag on for months or even years, and that a favorable decision was doubtful. The threat of a competing line, however, was immediate and certain.[84] But there was "one trump-card still in Mr. F. O. J. Smith's hands," observed Craig; namely, his absolute control of 350 miles of telegraph between the news and the market. On the last day of December 1849, Smith dropped his bombshell. He informed the AP that he would no longer suffer Craig's messages to pass over his lines.[85]

Craig later related that after Smith's notice he hastened to New York to tender his resignation to the executive committee of the NYAP. In New York, however, he was greeted by the good news that O'Rielly had just completed his rival line between Boston and New York. The situation had thereby been altered dramatically, and the association declined Craig's resignation.[86]

This made for a good story, if not an accurate one. O'Rielly had completed his line in September, and by December 1849, it must have been ready to transact business.[87] Yet Smith did have exclusive control of some 110 miles of wire between Boston and Portland. Craig and the AP controlled the news at source and held the terminus of the telegraph, in Halifax. A monopoly in time over the foreign news, however, necessitated domination of all stages in the process of news gathering and transmission.

To defeat Smith, the AP would ultimately have to break his telegraphic monopoly between Boston and Portland. They appealed to O'Rielly, the monopoly buster, asking him to extend his Boston line to Portland. At this juncture, they were apparently even willing to invest directly in such a line. O'Rielly, however, was satisfied with a guarantee that the AP would employ his line exclusively, and he vigorously pursued the undertaking.[88]

Until the line could be completed, the AP had to ensure its supremacy in the foreign news field despite 110 long miles of telegraph that was off-limits to them. Craig immediately went about arranging a special locomotive express to run between Portland and Boston with each dispatch of foreign news.[89] The considerable financial resources of the New York newspapers made the heavy price of such an express virtually insignificant. More troubling, on the face of it, was the three- to four-hour delay involved in running a locomotive express against the lightning line. The transmission of Craig's 3,000-word news dispatch from Halifax usually took three hours or less. A news agent friendly to Smith could easily send a news report from Halifax to Boston while the locomotive was chugging along the road from Portland. But the

NYAP's status as a preferred customer over the provincial lines proved to be the solution. Accepting the NYAP as representing the masses in the contest to control information and recognizing the force of a news consumer who could guarantee payment for 3,000 words under all conditions, the commissioners of the Nova Scotia telegraph, in full accord with public opinion and with the concurrence of the House of Assembly, stopped all competing dispatches of foreign news until Craig's railroad express reached Boston.[90]

Although Smith was clearly beaten, he pushed on recklessly, apparently motivated by sheer rage and vengeance. Rather than retreating from his ineffective nonintercourse with the AP, on 20 March 1850 he expanded the ban to include not only Craig's press reports but any and all messages to or from the eastern lines.[91] Moreover, realizing that O'Rielly's line was soon to connect Boston with Portland, he renewed his attempts to acquire enough stock in the Maine line, running from Portland to Calais, to control an additional link between Halifax and New York.[92] Smith owned a large block of shares in the Maine line, part of which represented his quarter share of the patentees' stock, and most of which he subscribed for upon the establishment of the company.[93] Yet Smith was short of a majority. He therefore approached Morse, through his agent Kendall, offering to purchase the three-sixteenths of the stock that Morse had received for his share of the patent rights. Smith thus expected to acquire a majority and to extend what was now a war of attrition with the AP.

Kendall and Morse, up to this point, had been surprisingly quiet. Notwithstanding their enmity toward Smith, they had not joined in the choir of condemnation ringing through the country's press. They well knew that Smith was fighting Craig and the AP in order to defeat O'Rielly and his plan of breaking the monopoly created by their patent. The New York-Boston route, the "telegraphic race course of the country," was merely the test case that would determine the effectiveness of O'Rielly's threat.

Initially, Kendall was apparently ready to aid Smith by turning over to him Morse's stock in the Maine line, stretching from Portland to Calais. Even in this emergency, however, Smith could not control his proclivity to drive a hard bargain. He declined every proposal Kendall offered; and with his rapacity met by Kendall's notorious obstinacy, negotiations stalled. Once Kendall and Morse became convinced that Smith had lost the war to prevent competition along the eastern seaboard, they did a quick about-face. They withdrew their proposal to sell their stock in the Maine line, jumped on the bandwagon of public opinion, and joined the chorus of jeers against Smith. They made no secret of the main reason for their decision to abandon Smith. The contest between the Morse patent and "the pirates" was only beginning. The other patentees could not afford to be linked to Smith, in case O'Rielly decided to move against their lines. Because the hopes of the Morse patentees now rested entirely on legal suits against O'Rielly, there was danger that public opinion would sway the courts. For the patentees, Smith's moves were now "fatally injurious to all interests connected with the Morse patent."[94]

The press, meanwhile, was having a field day condemning the telegraph monopoly, and the worried Kendall wrote Smith:

> Although I hope and believe that the firmness of our courts will shield our rights in Morse's patents from the disastrous influences of an adverse public opinion, which is daily increasing in volume and force, is there no danger on this score? . . . Will not every bias of [judges' and juries'] minds be against you? And is there no danger of its having, unconsciously to themselves, a most disastrous effect on their decision?[95]

Nor could Morse and Kendall afford to alienate the AP. Opposition telegraph lines were expected to spring up parallel to their own, and the Associated Press's position as a dominant customer made them wary of that organization's shift of patronage.

When the contest ended, there were several victors and one big loser. "The project of a general monopoly in the brains of Fog Smith," the *Herald* commented, was as "completely scattered to the winds as any mist that ever hung over the seaboard."[96] Smith had only the satisfaction of never having to repeal his nonintercourse order. O'Rielly's line to Portland was completed by the fall of 1850, and all news reports from the eastern lines were transmitted over it and over O'Rielly's "Merchants' Line" from Boston to New York. The press no longer wished to use Smith's line.

Henry O'Rielly emerged as the public's savior from the oppressions of a powerful monopoly. He proved the effectiveness of the Bain telegraph and of competition in the telegraph industry. Planning to expand his system along every important route on which the Morse patentees operated, he hoped, prematurely perhaps, that he had secured the loyalty of the telegraph's greatest single customer, the NYAP. The NYAP was also, though less conspicuously, a winner. It had to make some concessions to retain its exclusive privileges over the Nova Scotia line; for example, it had to share its foreign news with any newspaper that was willing to pay a fair price. As will be demonstrated, this might have been a blessing in disguise.

Craig was the all-around winner. Smith's various schemes provided

Craig with an opportunity to demonstrate his considerable talents. More important perhaps, Smith's heroics gave Craig some useful insights into managing a monopoly of news. He emerged from the contest with a deeper understanding of the significance of control over the telegraph wires, as well as over the market for news.[97] Craig also secured, in the aftermath of the contest, a formidable position from which to exercise such control. Late in 1850, the NYAP decided to replace its general telegraphic agent. There was little doubt in New York as to who would best fill the position. Daniel H. Craig, the agent for the foreign news, was called from the battlefields of New Brunswick and Nova Scotia to prepare for the new job. When on 18 May 1851 Alexander Jones formally stepped down from his position as the general agent of the AP, Craig took his place. It was in this position that he brought to fruition the valuable lessons he had learned about the formation of monopolies of information.

5

"A Monstrous Monopoly Is Created": The New York Associated Press Goes National

When Craig was promoted, he left behind him an extremely effective machinery to manage and control the foreign news.[1] Moving to New York as the association's new general agent, he found its domestic operations "an utterly contemptible affair." They were, reminisced Craig in 1866, "without system, head or heart—a rickety disjointed affair in every respect."[2] At that time, late in 1850, "there was no cooperation [with the Associated Press] on the part of a very large proportion of the daily journals of the country, and [the AP] was without any revenue from outside papers whatever, except about $100 per week, for European news via Halifax."[3] The first item on Craig's agenda as general telegraphic agent was to bring the unaffiliated newspapers over to the AP. He would thus transform the association from a joint news-collecting operation serving the major New York dailies into the major distributor of telegraphic news to the American press.[4]

Control of the foreign news was the key to making the AP the dominant supplier of the nation's news. Newspapers "could not do without the earliest foreign news," observed a telegraph leader, "and they could not get it except from Craig."[5] Craig thus had irresistible bait with which to lure American editors into cooperating with the NYAP. Meanwhile, the association realized that to maintain its status as the preferred client of the Nova Scotia line it would have to prevent any other news supplier from claiming the position of legitimate representative of the American Press.

At the time, many out-of-town papers received foreign news through Carr and Johnson, proprietors of the United States Newspaper Agency. From the beginning of the Halifax telegraph service, these agents supplied their client newspapers with European news collected by the NYAP, paying the association $200 on the arrival of each news-bearing steamer. Carr and Johnson provided telegraphic news mainly to the press south and west of New York, including the leading newspapers of Philadelphia, Baltimore, Charleston, New Orleans, Pittsburgh, Cincinnati, Louisville, and St. Louis.[6]

Toward the end of 1850, the NYAP canceled this arrangement. Carr and Johnson's appeal to the Nova Scotia telegraph authorities to be placed "on the same footing as the New York Associated Press" was rejected. The telegraph commissioners declared that because each newspaper they represented could get the foreign news directly from the NYAP, they were satisfied with their existing arrangement. The commissioners obviously had no desire to go through another lively struggle between competing news agents.[7]

The NYAP did not have to terminate any arrangement with its local competitors, the telegraphic news agents Abbot and Winans, who had made the mistake of siding with the two Smiths in their contest with Craig.[8] Now they had to be satisfied with supplying the Boston evening newspapers with short telegraphic reports from New York. On the arrival of steamers in Halifax, their Boston clients had to copy their news from the morning papers loyal to Craig. The vanquished evening papers of Boston even had to give credit to their rivals for the steamer reports.[9]

To strengthen the hold of the NYAP on the foreign news, Craig worked to secure first receipt of news arriving on ships reaching the port of New York. Here his carrier pigeons came in handy. Until then, the association's marine news arm, organized separately as the Harbor News Association, had employed boats to meet incoming ships and intercept their news. Craig improved the procedure by equipping the marine agents with carrier pigeons and having them fly the news from their boats to the association's headquarters. Later, Craig established a telegraph line from Sandy Hook to New York. His birds were retained, however, to transmit the news from the boats to the telegraph office at Sandy Hook.[10]

With firm control over the foreign news, Craig set out to "bring all of the leading Presses of the country into one general telegraphic news scheme." Before officially starting his work as the "European and domestic news agent of the NYAP," he went on a tour of the country's leading cities. Sounding out editors from Halifax to New Orleans, Craig was not surprised to learn that there was almost universal dissatisfaction with the existing arrangements. He found that editors throughout the country, "almost without exception, regarded their expenditures for telegraphic news much in the same light as they would regard a contribution to an unjust and unlawful tax-gatherer—a species of imposition which is not quite safe to resist."[11]

Craig diagnosed the problem correctly as "the natural result of a *bad system*—or rather, the want of all system" for the gathering and transmission of telegraphic news. Individual editors were helpless; the problem was "beyond the power of any one establishment to remedy"; and "what is everybody's business is nobody's." Clearly, there was no proper institution to mediate between the newspapers' demand for timely news and the telegraph system's potential to supply it. Craig tried to convince editors that there was in fact one establishment that could remedy the situation: the NYAP. He promised them he would "bring about an entire reform" and develop the proper organization to bridge the institutional gap between the press and the telegraph.[12]

Craig went on to prepare a blueprint for the operation of an efficient system of telegraphic news gathering and news flow that would encompass the whole country. The plan was intended to achieve "a uniformity and concert of action . . . between all of the presses connected in our proposed arrangement."

> All shall contribute, in proportion to means and relative advantages to be derived, to the expense and trouble of collecting and transmitting, from one end of the Union to the other, all important news—as well that relating to commerce as to general events—the wish being to raise the standard of telegraphic reports, both as regards the *matter* and the *manner* of the same—to make them what they ought to be—*reliable for accuracy,* and the medium through which all *really important* or decidedly interesting news shall be placed before the public, with the utmost despatch.[13]

This plan, issued in May 1851 when Craig officially assumed his management responsibilities, required some four years to be fully implemented. Once the great majority of newspapers were enlisted, it proved exceptionally resilient. The NYAP operated according to the original features of Craig's plan almost to the end of the nineteenth century. Even the few competing organizations that emerged during the course of the century found no better system and adopted the main characteristics of Craig's scheme with only minor modifications.[14]

Craig's plan called for the foreign news to remain the exclusive domain of the NYAP. Domestic news, of either commercial or general interest, was to be collated by agents the association placed in the nation's important urban centers. These agents were not expected to function as journalists in the relatively new capacity of reporters—those collecting the crude news in the field—rather, they were to act as intermediaries between the newspapers in each agent's jurisdiction and AP headquarters in New York.[15] The local newspapers were expected to collect the primary news and supply it to the agent. The agents in turn sent the news over the wires to Craig's desk at AP headquarters. They were also, as far as possible, to "report to each other." Then it was the agents' responsibility to supply the consolidated report prepared by Craig's staff at the national news emporium to the newspapers in their respective areas.[16] The association's national system of news distribution thus yielded a comprehensive domestic news-gathering network serving the New York members.[17]

The success of Craig's plan depended on the number of newspapers he could incorporate into his arrangement. His primary objective was to secure for his association the status of legitimate representative of the American press, which could best be reflected in the number of newspapers it actually supplied with news. There was also a simple economic reason the number of newspapers served was important. The cost of gathering the news and the overhead expenses of running the news organization were constant; so the more newspapers that joined the service, the smaller the cost to each of the New York members. The main expense in the operation of a wire service was, however, the cost of transmitting the telegraphic news report from New York. Here too, the more newspapers that received the report at each point, the cheaper the cost of transmission per client.[18]

No less important than the scale of the service was its scope, or the geographical distribution of the newspapers that Craig could bring over to his arrangement. Indeed, the spatial comprehensiveness of the membership would determine the completeness of the association's news coverage. Moreover, the telegraph companies arranged special circuits for transmitting the consolidated news report from New York. They tried to build far-reaching circuits that would allow the report to reach as many destinations as possible in one "writing." When transmitting an AP report to New Orleans, for instance, the report could be dropped at Philadelphia, Baltimore, Washington, Charleston, and all other stations along the line, at the cost of the paper on which the otherwise idle telegraphers took down the message. Because what the telegraph companies sold was the occupation of their circuits, when the distribution of customers made for many, long circuits, there was an exponential rise in the number of potential customers to whom the news could be supplied with minimal added cost, and therefore offered very cheaply.

The economies of scale and scope were thus virtually unlimited. The more customers, the cheaper the product and the greater the revenue, with no decline in the marginal profit. The more widely distributed the customers, the better and cheaper the service. And whatever made expansiveness effective under these conditions made competition ineffective. Put together, this meant that the wire service business was a natural monopoly. No one understood this better than Craig.[19]

To make a nationwide system work, it was necessary to win over to the service a certain critical mass of newspapers. The peculiar dynamics of the system would make the assimilation of the rest of the newspapers almost inevitable. Although it took some four years for that to happen, there was never any serious doubt about the outcome. The investment necessary to carry the news organization along before the economies of scope and scale began to express themselves was gladly borne by the affluent New York members. Indeed, they would willingly have carried the full cost of procuring the news anyway.

The one potentially serious challenge to Craig's plan of establishing a national monopoly of news supply came from the telegraph companies. Craig's system was predicated on speedy and comprehensive news gathering, on transmitting it rapidly and cheaply, and on an efficiently managed and well-coordinated system to perform these tasks on a national scale. The telegraph companies had a decided advantage over Craig in the most expensive aspect of the scheme, the telegraphic transmission of the news.

Establishing a coherent system to operate nationally may have appeared at the time beyond the power of telegraph companies. In what a historian of the telegraph called its "era of methodless enthusiasm," the industry was a chaotic collection of small, independent companies. Yet the first signs of change were beginning to appear. At the time Craig was introducing his organization for the management of a national news system, the telegraph was entering what the same historian considered its "era of consolidation."[20]

From bitter experience, Craig knew how dependent the news business was on the telegraph. Moreover, he must have understood that the splintered state of American telegraphy was only temporary and that the industry had some pronounced characteristics of a natural monopoly, as the news service did. If Craig needed a practical illustration of the monopolistic nature of the telegraph industry, he was soon provided with one in his own backyard. Nowhere was the trend toward consolidation more apparent than on the New York-Boston route.

With great satisfaction, Marshall Lefferts, president of the New York and New England Telegraph Company (known as O'Rielly's "Merchants'" or "Bain" line), informed investors late in 1850 that the directors had decided on a 6 percent dividend from the profits of the line's first half-year of operation. The AP paid considerably for the transmission of its news, private business was booming, and the company's prospects appeared ever so bright.[21] Yet the following year was extremely disappointing. No dividend was paid, and it was felt that there would be no buyers for an issue of additional stock deemed necessary to finance operations. Fog Smith and his veteran New York and Boston Magnetic Telegraph Company entered into a vicious price war, trying to exhaust the thin resources of O'Rielly's over-expanded group. The Bain line had no choice but to "meet Mr. Smith upon his own ground, and with his own weapons."[22] The price of transmitting a ten-word message between New York and Boston dropped from a peak of fifty cents in the days of Smith's unopposed control down to ten cents.

Those with interests in the Bain line had other worries, too. A federal court in Philadelphia, unswayed by public opinion, ruled that the Bain instruments used on a line connecting New York and Washington were an infringement on Morse's patent. With that courageous decision as a precedent, the prospects of the Boston Bain line's winning the suit that Smith had brought against it were poor.[23] A legal defeat would make their investment worth the price of two 230-mile lengths of used wire and some tons of scattered timber. In contrast, Smith had a valid patent but not much of a telegraph line. His line, by losing the patronage of the AP and many private customers to its competition, and by indulging in a reckless price war, had insufficient income to cover operating expenses, let alone be maintained properly. Its condition rapidly deteriorated, and when put up for sale early in 1852, no serious offers were made.[24]

Consolidation of O'Rielly's and Smith's operations was inevitable. On 1 July 1852, O'Rielly's New York and New England Telegraph Company was merged with—or, as Craig saw it, was absorbed by—Fog Smith's New York and Boston Magnetic Telegraph Company. Together they formed the New York and Boston Union Telegraph Company.[25] There was now only one reliable telegraph operating between those two cities.

Surprisingly, the consolidation took place without causing the slightest ripple in the amicable relations between telegraph and press. For once Smith was not looking for trouble; all he allegedly wanted was "to forgive and forget and make money."[26] "I have long since ceased to feel any uneasiness towards the Associated Press," he wrote two months after the merger. The consolidation came when his line was on the verge of disaster, and he could not afford to risk it before the final arrangements had been made. Besides, Smith was still seriously negotiating the sale of his line, and discord with the press would reduce its value. Under the circumstances, to use his new-found power to "disturb the New York Press" he considered "both madness and folly."[27]

Nor could Smith, had he wanted to, do much to interfere with the press business. One of the conditions of the merger was that the Bain people would control the management and practical operation of the Union line. The president chosen for the new company was an associate of O'Rielly's and the three-man committee charged with press arrangements was headed by Marshall Lefferts, Craig's ally, and included another Bain man.

The AP appeared to be in just as conciliatory a mood as Smith. "This is the end of the war," announced the *New York Courier and Enquirer* on the occasion of the consolidation. "We have no harsh feelings towards that officer [Smith] who has certainly shown that he possesses the quality of *perseverance* in a degree rarely manifested by any thing that goes on two legs." Nor would it help the association to show displeasure or dwell on the "many sleepless nights" Smith had caused it.[28] The AP had no choice but to use Smith's line, and there was no reason not to approach the situation positively.

Craig, however, had no intention of submitting himself and the AP "a second time to the tender mercies of Smith," and he at once set to work to get out of the predicament.[29] He found an alternative to the consolidated line in the cheaply constructed, poorly maintained, and badly managed House line between Boston and New York, which was "at its last gasp." Craig decided there was no reason it should so remain. He began sorting out the intricacies of the ownership and claims on the faltering line, with the intention of acquiring control.[30]

Rumors of Craig's interest in the House line gave the Bain party a wonderful excuse to put pressure on Smith.[31] They demanded control of the management of the Union line for at least five years, to reassure the press that its interests were completely safe in the Union Company's friendly hands. The Bain people, as a further inducement and as a demonstration of their friendliness, began harassing the association of the Boston evening papers. Users of the Morse line and customers of Abbot and Winans, these patrons were part of Smith's endowment to the consolidation.[32] In no position to see the merger fall through, Smith capitulated and granted the Bain party the management of the concern for five years. He even assented in principle to give up his controlling stock just to reassure the press. It would take Smith a while to digest "the savory dish set before him by . . . McKesson and others of the Bain line," noticed an amused Craig.[33]

In September, after all the groundwork was laid, the Union company offered the AP a five-year contract for conducting press business under extremely favorable terms.[34] The association, however, had by then an alternative proposal, for Craig was perfecting his arrangements for a takeover of the House line. He secured an option to purchase it for

$40,000 in cash and the cooperation of the efficient John McKinney, former superintendent of the Bain line, for putting the line in good working order. Craig now offered the independent line to the NYAP.

Although the idea of the press's taking an interest in the ownership of telegraph companies was not new, editors had not previously thought of becoming actively involved in managing and operating telegraphs. In 1849, the *New York Sun* announced to the world that it was establishing its own telegraph system from Boston to Washington. "The telegraph system in this country is nothing but an odious monopoly and must be broken up," the editor declared. "The most important use of the telegraph is its service to the press and to the press it properly belongs, as we shall shortly show by a practical demonstration."[35] The *Sun*'s new telegraph system never materialized.

At the height of Smith's unopposed power, the AP deliberated for the first time whether the association as such should become financially involved in telegraphy. The editors were considering a joint investment in the line that O'Rielly was contemplating between New York and Boston. After a thoroughgoing discussion, they decided not to pursue investment. The reasons for this important policy decision were never given. It is likely, however, that the editors viewed the joint ownership and management of an enterprise that was so far removed from their main line of business as impractical. Indeed, the members of the AP in a meeting of 28 August 1851 declined Ezra Cornell's request for a loan to help complete his western system, on the grounds that "matters of this kind" were "beyond their powers as an association."[36] Such a venture would appear all the more difficult when considered in the light of contemporary conventions of business management. Traditionally, American business enterprises were proprietor-managed. Only with the growth in the size of ventures, and with the consequent diffusion of ownership and specialization of operations, did a professional management begin to assume responsibility for directing affairs.[37]

It was only gradually that the NYAP began disassociating management from proprietorship. When the NYAP established the Harbor News Association, its management rotated among the five partners, with each newspaper assuming responsibility for one month in turn.[38] In the early stages of common telegraphic news gathering, responsibility for procuring the foreign news also apparently rotated among the members. Later, it was entrusted to the *Boston Daily Mail*, and only subsequently was the responsibility for the foreign news delegated to an agent. Even the association's joint national news procuring arrangement was managed at first out of house, by Alexander Jones's independent agency. When Craig established a large-scale national news service on behalf of the AP, the move reflected an important policy change.

The partnership was now prepared to assume joint responsibility for an operation that was run by a professional manager, independent of the member newspapers.

Presented with Craig's proposal that they purchase the House line, the proprietors of the AP once again considered the issue of management of telegraph lines by the press. The editors had to choose between two contrasting options. The Union line offered them extremely favorable terms, as well as permanence.[39] Craig suggested that the association take its telegraphic destiny into its own hands and control the medium for its news supply. Each proposition suggested a way of coping with telegraph consolidation; one by close alliance with, the other by control of, the telegraph. The editors, realizing they were faced with a serious situation, chose a third course.

In two long meetings of the full membership the editors convinced themselves that the ownership of telegraphic interests was too great a change. They declined Craig's offer of the House line on the grounds that "telegraphy was outside the business of newspaper editors." Yet the editors declined the generous offer of the Union line as well. They solved the dilemma by encouraging Craig to undertake the telegraph enterprise as a personal investment. The AP would secure the investment by giving its business exclusively to the House line. Craig, together with his friends James Spear and John McKinney, exercised the option they took for a controlling interest in the House line. By November 1852, McKinney succeeded in getting the line into good working condition, and it began transmitting the press news.[40]

The NYAP emerged from the crisis with an established policy regarding the trend toward consolidation in the telegraph industry, which it announced in one long sentence: "We think that the public require two distinct lines of telegraph; that competition in the telegraphic business is necessary to the community; that the Press, as an agent of the public, and as a part of its duty as such, should aid in the support of *independent lines;* and that the interests of the Press call for the organization and public aid of such lines."[41]

Although competition prevented consolidated telegraph companies from taking over the news business, it also made room for alternative news agencies to emerge. Telegraph lines that did not benefit from the patronage of the AP would be glad to transmit for a competing news agency at low rates.[42] It was thus important for the AP to keep on good terms with both sides. Therefore the association, avowedly to sustain competition, but also to keep everyone happy, decided in principle to carefully divide its patronage whenever two companies competed on the same route. Yet the policy proved quite flexible, and it was managed pragmatically. As in the case of the competition on the New York-Boston route, when it was deemed necessary to support one telegraph company over another, the press patronage shifted accordingly.[43]

As a further precaution, rather than bargaining for a cheap price, the association as a rule agreed to pay the highest prices the telegraph companies asked, with the important condition that no other news messages would be sent at a lower rate. The association was even willing to give telegraph companies a share of the actual news business. It granted them the right to sell the association's reports to the press of all but the major centers along their routes without cost.[44] This policy, besides keeping the telegraph companies happy with serving the AP and saving the association much administrative work, had an additional objective in view. The more newspapers that were supplied with the AP's news, the less the opportunity for a competing news agency to establish itself.

The divide-and-rule policy the NYAP adopted was intended as a sufficient response to the trend toward consolidation in the telegraph industry. Craig, however, sensed that the association's leverage was not in itself enough to counteract the mounting pressures within the industry toward oligopoly. He was certain that the telegraph would ultimately take the form of a national monopoly. Craig also understood that his plan for a centralized, systematized arrangement for collecting and transmitting the nation's news depended on a unified national telegraphic network. At the time, Craig's single message from Halifax to New York, when European steamers arrived, had to pass over the wires of five different companies, each employing different procedures, separate accounts, and varying policies. Coordinating the smooth transition of the message, even between two neighboring companies as friendly to each other as the Nova Scotia and the New Brunswick lines, involved what Craig would subsequently describe as "numerous details, sudden emergencies, conflicting interests and ever changing phases." The welfare of a national news system clearly depended on a nationally integrated telegraph network.[45]

Therefore, in 1853 Craig apparently presented to the editors a revolutionary plan. He made a "vigorous effort . . . to induce the Associated Press to buy up the direct trunk lines between Boston and New Orleans, on the seaboard, and the great cities of the West and Southwest, which could then have been done for less than $1,000,000." The editors in conclave "declined to permit [him] to go on with the enterprise." "Had they assented to my wishes," offered Craig in retrospect, "the Western Union Telegraph Company would have been buried in its infancy," and the telegraph monopoly "would have been as a tail to the Associated Press kite."[46] The AP may have declined the proposal; yet Craig, for his part, had a personal aversion to flying as a tail to anyone's kite.

Late in 1856 the NYAP reversed its earlier decision to divide its business in order to encourage competition in telegraphy. It now tried to concentrate its patronage as much as possible on one telegraph concern in the East—the American Telegraph Company; and one in the West—the Western Union Telegraph Company.

The American Telegraph Company was a newcomer, formed only the previous year. The prominent contemporary telegrapher James D. Reid understood that the company "had its origins largely in a press movement, which more or less affected, during many subsequent years, its policy and success."[47] It was Daniel Craig who led the press movement that gave rise to the American. The company was, as Reid suggested, "the natural outgrowth of the control of the line between New York and Boston by McKinney and Craig."[48]

Craig had a grand vision for the future of the line he owned in the interest of the AP.[49] He believed it would serve as the foundation for a telegraph monopoly that would advance the interests of the wire service. This telegraph monopoly would relieve the AP of the need to maneuver between competing lines and from becoming a tail to the kite of an unfriendly telegraph monopoly.

The project at first appeared ridiculously farfetched. Even Craig realized that an interest of $40,000, albeit on a key route, was not a sufficient lever to turn on its head an industry now capitalized in millions. Yet when he ventured to establish the telegraphic foundations for the AP news monopoly, some heavyweights were entering the telegraph business for the first time. It was in collaboration with New York capitalists, of the caliber of Peter Cooper, Abram Hewitt, Moses Taylor, Wilson G. Hunt, and their leading spirit, Cyrus W. Field, that Craig set out to make things lively in seaboard telegraph affairs.

Field, a prosperous New York paper merchant in early retirement, became involved in telegraphy through the influence of Frederick N. Gisborne, a Canadian. Former superintendent of the Nova Scotia government line, Gisborne in 1852 organized a company chartered by the Newfoundland legislature to establish a telegraph line across the province. This line was to connect to the mainland by means of a submarine cable. Having exhausted his resources before completing the project, Gisborne tried to interest New York capitalists in the venture. Cyrus Field responded to the idea of extending the telegraph to the easternmost extremities of the North American continent with great enthusiasm. He realized that by erecting a line across Newfoundland, a number of days could be shaved off communications with Europe. Whether it was Gisborne or Field who first recognized the value of the Newfoundland line as a link in a trans-Atlantic cable is moot. It was Field, however, who began pushing the project of a trans-Atlantic cable in earnest, and he enlisted friends among the leading New York capitalists to invest in the Newfoundland line and in the cable venture. The lion's share of the bill for the project was, however, footed by British investors.[50]

It did not take long for Cyrus Field and his group to learn what Craig already knew: that receiving early advices from Europe was worthless without command of reliable channels to communicate them to their American destinations. To prevent potential interference with the cable dispatches once they reached North America and to reap the full profits of their transmission, the cable entrepreneurs would have to control a smooth system of inland routes.

An alliance in domestic telegraphy between the cable interests on the one hand and the wire service interests on the other was only natural. The foreign news was an important element in the power of the AP. By cooperating with Field's group, the association could secure control of the foreign news over the Newfoundland line and one day, perhaps, over an Atlantic cable. What Craig and the AP had to offer, for their part, was a considerable proportion of telegraphic patronage—national and international—and public support for the cable entrepreneurs.

Craig's personal contribution to this combination of interests was crucial. By the time Field's group began to consider the problem of connecting the cable with the urban centers along the seaboard, Craig had formed an alliance with James Eddy, superintendent and major stockholder in the Maine Telegraph Company. Together, they controlled a continuous line from New Brunswick to New York, as well as several important interior lines.[51] Besides, Craig had great leverage in the Nova Scotia and New Brunswick telegraph companies, as their prosperity was totally dependent on his press business. Thus, he could assure the cable entrepreneurs control of a continuous channel of communications from the mouth of the cable on to New York.

But Craig's most important contribution to the undertaking was discovering David E. Hughes, a music professor from Kentucky who claimed to have invented a telegraphic instrument that was wholly distinct from those of Morse, Bain, or House. Craig immediately purchased the rights to the Hughes invention. He realized that possession of the promising instrument would make it possible for a new concern to build telegraph lines on any route it chose. It was a key to breaking the monopoly of the established telegraph companies.[52]

Cyrus Field's group joined with Craig and his associates to form the American Telegraph Company on 1 November 1855. The company was capitalized at $200,000 and incorporated under the New York telegraph law—the one state law that recognized priority for press dispatches.[53] On the same day that the company was organized, Craig sold the rights to the Hughes instrument (eventually patented in 1857) to his partners in the American Telegraph Company. Shortly thereafter, the Maine Telegraph Company leased its lines to the new company; a lease of the New Brunswick line followed; and on 30 May 1856, Craig's House line sold out to the American.[54] The patronage and support of the AP were now mobilized to advance the interests of this emerging "titan of the East."

It was not long before the AP found it politic to concentrate its western patronage as well. At the same time that the northeastern telegraph companies were merging into the American Telegraph Company, a similar movement was under way in the West. The New York and Mississippi Valley Printing Telegraph Company, a Rochester concern, began acquiring small lines in the Midwest and by 1856, as the Western Union Telegraph Company, it emerged as the dominant power in western telegraphy. The policy of encouraging competition in that region was no longer realistic or safe for the AP. Accordingly, toward the close of 1856, the association entered into an agreement with the friendly "colossus of the west." The AP made the Western Union "the contractor for the transmission of all news reports from and to Philadelphia and the west and from and to Buffalo and the west." In return for its exclusive use of Western Union wires, the AP was secured from competition in that region, and it received among other privileges "the facilities of all the wires of the W. U. Co. for the interchange of news."[55]

The new conditions in the telegraph industry made it necessary for the AP to overhaul its internal arrangements and its relations with its client newspapers. On 21 October 1856 the NYAP reorganized, styling itself the General News Association of the City of New York. The reorganization was undertaken to aid its telegraph allies in their bitter struggle with the established telegraph companies controlled by the Morse patentees.[56] But no less important, the reorganization represented an attempt to use the consolidation in the telegraph industry to fortify the position of the AP in the news business.

The new rules established by the General News Association operated in the main to centralize the telegraphic news business. Recognizing that an all-out competition in telegraphic "specials" between the affluent New York City dailies was dangerous to them all, the association decreed that all members had the right to use the specials of any other daily, after paying their share of transmission expenses. As specials could be neither exclusive nor competitive, the incentive to reach out and procure them was greatly diminished. The common machinery of the Associated Press came to dominate the telegraphic news gathering for the member newspapers.[57]

Thus, a decade of cooperation proved that there was room enough for newspaper competition without involving rivalry for telegraphic news. Yet the rules allowed exclusive specials from Washington and Albany, from political conventions, and in several carefully enumerated circumstances. The member newspapers were thus allowed some freedom, and readers some diversity, when it came to purely political reporting, even if it was transmitted by telegraph.

The most important departure in the new rules was in making the AP the exclusive medium for procuring telegraphic news for the client papers. Clients were not even allowed to use the telegraph independently to forward dispatches by reporters they sent out to cover news events on their own behalf. This rule, "forbidding the reception of news through any other channel than D. H. Craig," was viewed by subsequent opponents as "calculated to create and perpetuate the most odious monopoly that the wit of man could devise."[58] The regulation, besides preventing potential competition and strengthening the hand of the association's allies in the telegraph industry, also reflected the efficiency and comprehensiveness of the system Craig had built. There was nothing important a newspaper could procure telegraphically that Craig's organization did not supply, and no competing organization could provide better service. Craig had no doubt what newspapers, presented with the choice of taking his report exclusively or doing without it, would decide. Indeed, newspapers throughout the country almost without exception found that they were best served by the NYAP and submitted to the new rule. Thus, five years after Craig inaugurated his system, it became powerful enough to legislate itself as the nation's telegraphic news monopoly.

Some perceptive observers pointed out the far-reaching significance of the centralization and standardization of news throughout the country. The most obvious characteristic of a monopoly of news was its potential for abuse. Fog Smith was an expert on that aspect of the problem: "The mythological power attributed to Jupiter was not greater than that of the man at the head of the Associated Press; his power came nearer to that of omnipotence than anything else within [my] knowledge; he had more power to make and unmake presidents than either party; he could send or withhold such news as he chose, and thus shape public sentiment at will."[59] Similarly, a group of telegraph leaders complained that "through his subordinates [Craig] can regulate the tone of the news despatches; he is editor in chief of the Press of the whole country."[60]

Craig "had a vocabulary of wonderful uniqueness and force," observed a contemporary, adding that "he was always himself." Craig's published response to these statements therefore came as no surprise. "I am only sorry," he informed the public, "that my power is not as omnipotent as the committee and Fog Smith assert, for, if it were, I would summarily string up by the ears and suspend from the Telegraph building in Wall street, this whole brood of hypocrites."[61] In response to similar accusations, Craig, in a more contemplative mood, pointed out a simple truth: "In the nature of things, there must be somebody to whose care the telegraphic news of the country must be confined for the convenience of the press."[62] Craig thought that nobody could do the job better or more honestly than himself, and obviously it had never occurred to him that the telegraphic news business could take any other form than that of a national monopoly.

Not long after the formation of the American Telegraph Company, the AP harvested the first fruits of its cooperation with the cable interests in the foreign news field. In the summer of 1855 Cyrus Field attempted to complete the Newfoundland line. A storm prevented the landing of a cable from Cape Breton to Cape Ray on the first try, and only after a second, expensive expedition was the mission accomplished.[63] To Field, the laying of the cable connecting Nova Scotia with Newfoundland was merely a preliminary step. The good people of Newfoundland were not expected to cover the investment of $800,000 with their private messages to the mainland, even if they used the cable to the end of time. For telegraphers, the line had no value besides providing a connection for a future cable to Ireland.[64]

To Craig, the news-gatherer, and to the AP, the completion of the Newfoundland line was in itself a great victory. Following the best tradition of American news gathering, from the days when Samuel Topliff intercepted vessels in the offings of Boston Harbor with a rowboat, Craig positioned news yachts off Cape Race. By intercepting the mail steamers off the cape, Craig could get the news an average of two days before it arrived in Halifax. The Newfoundland line thus cut the time lag between Europe and America to a week or so. Swift vessels could even reach Southampton from Cape Race in a matter of five days.[65]

Although the line to St. John's, Newfoundland, was a giant leap for news gathering in general and for the AP in particular, Craig and others of the American Company knew their fortunes ultimately depended on the successful laying of the Atlantic cable. Field made an attempt to lay the cable in the summer of 1857. After it had been payed out over a quarter of the distance between the continents, the wire snapped, and hundreds of miles of wire—together with a small fortune—sank to the bottom along Maury's Plateau.[66]

In the summer of 1858 Field made a second attempt at laying the cable, undaunted by the great financial crisis of 1857 or by his failure the previous summer. Petitions and remonstrances from the Morse patentees to block government support for Field's project were of no avail, and in June of 1858 the United States Navy was well represented in the fleet that set out to lay the cable.[67] This attempt failed, as did another shortly thereafter, when the cable again snapped. In the middle of July, Cyrus Field once more attempted to lay his Atlantic cable.

The hopes and fears of many accompanied the small fleet that set out on the final expedition of 1858. Cyrus Field had laid his prestige, years of laborious efforts, and a fortune on the line. Morse, Kendall, and Fog Smith well knew that a successful cable would shatter their hopes of containing the competition of the American Telegraph Company along the seaboard. The AP looked on, realizing that should the cable be successfully laid, it would become the exclusive purveyor of a continuous flow of foreign news to the American press. Failure to lay a functional cable would seriously threaten its position in the national news business.

The AP had identified its telegraphic interests with the American Telegraph Company. The future of that company—in essence, a news-cable concern—was on the line. A successful cable would pave the way for the American to dominate telegraphy in the East; failure would strengthen the hand of its opponents—by now entrenched enemies of the AP—and perhaps even destroy the company. Americans from all walks of life—not just telegraphers and news-gatherers—took a keen interest in the fortunes of the cable expedition, hoping with Reverend Henry Ward Beecher for a "communion of the Old World with the New."[68]

On 5 August 1858, a telegram from Cyrus Field to the NYAP traversed the whole length of the American Telegraph Company's lines, from St. John's, Newfoundland, to New York City. It announced the successful laying of the Atlantic cable.[69] The headline of the *New York Times* the following morning proclaimed "The Great Event of the Age," and the press of the entire country followed suit. A week and a half later, when the AP reported that the queen of England and the president of the United States exchanged pleasantries by cable, the country was elated. Celebrations in New York reached such a pitch that City Hall, brilliantly illuminated by torches, fireworks, and rockets, caught fire.

Yet in the first days after the AP announced the good news, enthusiasm was restrained by fears that the notice might be a newspaper hoax. The editor of the *Hartford Times,* who "had considered the chances against success at least 100 to 1," found it hard to believe the news, yet he "could not discredit the dispatch which came through the reliable source of the Associated Press." He therefore sent a special telegram to Mr. Craig, "asking him if there could be any doubt as to the reliability of the news." Craig's immediate reply was: "It is true. You may bet your hat on't."[70]

The skepticism of the first few days after the AP announced the successful laying of the cable re-emerged some three weeks later. Cable authorities announced that they could no longer decipher the dim signals passing between the continents. The initial disbelief in the technological potential to establish trans-Atlantic telegraphic communications, combined with the admitted failure after such a short period, made many suspect that the Atlantic cable of 1858 was a "humbug." The cable, many believed, was either never laid, or if actually laid, never transmitted any legible signals.[71]

Several aspects of the performance of the short-lived cable fed those suspicions. One was that throughout its brief career the cable was not open to public use. With perhaps one exception, its facilities were limited to communications of the cable entrepreneurs, who occasionally provided the AP and the British newspapers with news items.[72] Nor did anyone besides the company's personnel ever get to visit or witness the North American terminus of the cable. Even the identity of the operator at Trinity Bay, Newfoundland, remained a mystery. His name was variously given as De Sauty or De Santy, and nobody inside or outside telegraphic circles seemed to have known him or anything about him before the summer of 1858.[73] "Is there a De Sauty . . . Or is he a mythus—ancient word for 'humbug'? " pondered the poet Oliver Wendell Holmes's fictional professor at his breakfast table. "Born of stream galvanic, with it he had perished! There is no De Sauty now [that] there is no current!" was the satirical answer he received.[74]

Yet the gravest suspicions related to the news reports purportedly transmitted by cable. Electricity was supposed to convey information, even over the greatest distances, instantaneously. What skeptics found striking was that of the cable messages distributed by the AP, not one provided unequivocal proof of instantaneous transmission. A case in point was the exchange of congratulations between the queen and the president that had raised the public's enthusiasm. At the time the official messages were exchanged, Queen Victoria was on a visit to Prussia. Her cablegram was undated, and its text was probably given to the officers of the Atlantic Telegraph in England either before or shortly after the cable was landed on 5 August 1858.[75] It was handed to the president by the AP and distributed to newspapers only on 15 August 1858.[76] The president's reply, given that day to the AP agent (who represented the cable company), was not published in the *Times* of London until 23 August.[77]

Virtually all the news items released by the cable authorities through the AP reflected news events that took place at least a week before they were reported to have arrived over the cable.[78] And newspaper readers knew that by intercepting steamers off Cape Race the AP could receive or transmit news, to or from Europe, in less than a week. The suspicion was that the AP collaborated with the cable managers to present steamer news and mail intercepted by the AP off Cape Race as having been received by cable.

When Morse demonstrated his telegraph in 1844, skepticism gave way to enthusiasm when numerous individuals could verify that Baltimore news reached Washington before any other means of communication could have conveyed it. In the case of the cable of 1858, such verification was impossible. The AP was the one source of early European news reaching America, whether by ship or by cable; and it had an obvious interest in using this exclusivity to establish the credibility of the cable.

The Atlantic Telegraph Company informed the public on 27 August that "the only news despatches thus far received through the Atlantic Cable have been addressed to the agent of the Associated Press, New York. No special or private despatches of European News have been passed over this line to any other address." "Observer," writing in the *Boston Courier* the following year, pointed out the significance of this fact:

> The associated press news arrangement[s], through which only one man [Craig] was announced as the sole recipient of news from Europe over the Cable . . . begin to stand out to the public comprehension for—*what they are worth!* . . . But for [these arrangements], all the presses of the country could not have been made . . . the instruments of this cool imposition upon the whole public of this Union, nor could the editorial corps of this whole country have been made . . . alike the dupes of this one agent at one and the same moment.[79]

Conclusive proof that the cable was a hoax was never produced.[80] Yet in one case the accusations that the AP collaborated with the cable interests to present steamer news as arriving over the cable are especially persuasive. The last purported cable dispatches published by the *London Times* were of 23 August 1858. Thereafter, the cable apparently showed no signs of life, yet the AP went on to publish detailed news dispatches, "by Atlantic Cable," from Valentia, Ireland, well after that date. On 28 August, it gave a relatively extensive news report of the British victory over the Gwalior insurgents in India and of the settlement of the Chinese question—news that was known in Paris and London on or before 20 August.[81] The news may well have arrived by the *North Star,* which left Havre on the 18th and was intercepted off Cape Race on 27 August. In fact, the full advices of the *North Star,* published after its arrival in New York port on 30 August, were surprisingly similar to the purported AP cable dispatch distributed three days earlier.[82]

The suspicion that the Atlantic Cable Company and the AP collaborated to mislead the public about the performance of the cable of 1858 found little expression in the American press. AP members probably found it impolitic, and client newspapers unsafe, to investigate the charges. I found only two exceptions to the general silence on the issue. In both cases, the newspapers involved were in revolt against the AP. They had nothing to lose from its ill-will and everything to gain from injuring its reputation.[83]

If Craig and the AP did in fact take part in a hoax, they had strong motives for doing so. By demonstrating that the cable was a partial success, their ally, Field, would find it easier to enlist investors for future attempts to lay another cable. It would also help Field and his friends unload some of his $400,000 worth of shares in the cable company, which previously no one seemed interested in taking off their hands, even at twenty cents on the dollar.[84] Furthermore, a large share of Field's stock was salable only on condition that a successful cable was laid. Indeed, even the charter for the cable company was conditioned on laying a successful cable by the end of 1858. And most important, perhaps, the more real the prospects of a future cable appeared, the stronger the position of the American Telegraph Company became, vis-à-vis its competitors along the Atlantic seaboard.[85] Because the AP was identified with the interests of the American and was assured of advantages in using the cable, the perceived feasibility of a future attempt strengthened its hegemony in the national news business.

To counter the advantages a successful cable might bring the American Telegraph Company, the Morse interests decided to pursue a cable option of their own. They were attracted to Taliaferro P. Schaffner's grandiose projection of a series of cables reaching Ireland via Iceland, Greenland, and Labrador. Contemplating the project, Smith and Kendall found that they were virtually checkmated by Field and his associates. The cautious New Yorkers were too shrewd to invest millions and leave the door open to competitors who could endanger their investment. Before embarking on the cable project, they had secured for their company the exclusive rights from the legislatures of Maine, New Brunswick, Nova Scotia, and Newfoundland to land cables off their shores. Massachusetts, however, was not immunized, and Field's competitors attempted to secure the right to lay cables off its shores. A petition was presented on their behalf to the Massachusetts legislature on 15 March 1859 for incorporating the Boston and Halifax Telegraph Cable Company.[86] The company's first project was the laying of a cable from Cape Ann to Yarmouth.

The leading spirits behind the initiative were Frederick N. Gisborne and Francis O. J. Smith, and they were warmly supported by the Nova Scotia Telegraph Company, by then a private concern. The company did not heed Craig's warning against placing too much confidence in the schemes of the "vile scoundrel trickster Fog Smith" and the "would-be swindler" Gisborne. The Nova Scotians neither shared nor appreciated Craig's prediction that ultimately "even the stupid people in Halifax will get their eyes open to the scoundraly conduct" of Gisborne and Smith, "bye and bye."[87]

Energetically pursuing the project, Smith realized that the public would not take kindly to an effort to injure Field after he had sacrificed millions in pioneering the cause of the cable—especially an effort by parties who did nothing but obstruct his work.[88] The project was therefore promoted as an effort to break up the monopoly of the NYAP over the national news supply.

The project received the support of a powerful group of respectable Boston merchants who had problems with Craig, with the Associated Press, and with New York. On 11 January 1859, they produced a strong resolution in support of the enterprise. They informed the public that "efforts have recently been made to consolidate all existing telegraphic companies in the United States into one company [the American]." The merchants had reason to believe that the American Telegraph Company "would place the transmission and distribution of the foreign, and much of the domestic news throughout the country, under the immediate control of the agent of the 'associated press' in New York."

The AP, they argued, would have "a concentration of power which . . . should never be confined to one individual, company or corporation." The association's agent, pointed out the Boston merchants, received the earliest knowledge of foreign news through the single line from Halifax. "May not that knowledge," asked the committee, "when received in advance of other messages to individuals, and which may contain the news of great changes in the markets, a declaration of war, or the conclusion of peace between nations, be used for individual and private advantage of an agent of the press, or that of his friends, to an extent not easily estimated?"

Yet the main complaint of the Boston merchants was that this one man operated out of New York. Fearing that "Boston . . . is to be made secondary in the receipt of foreign news," the committee called for support of the Halifax cable project.[89] The backers of the measure repeatedly claimed that the issue was a "general question between the city of New York and the eastern seaboard of Massachusetts" in the quest to control the first publication of news. They called for local patriots to support "this Massachusetts Bill."[90] Senators argued that "Boston was not treated fairly in the matter of telegraphic news in comparison to New York. Important news was frequently withheld from Boston merchants until after it was known in New York." This information was acted upon in the markets to the great injury of Boston merchants. "The gentlemen of the Associated Press took good care to look out for the interests of New York," stated one senator. And besides, he added, "Was not an effort constantly made to centralize everything at New York?"[91]

A majority of the state senators, however, revolted against having their vote on the bill serve as proof of their loyalty to Massachusetts.[92] They considered the bill on its own merits and found them rather slim. "Respectability and prejudice are to carry this measure through—nothing else is appealed to," complained one state senator.[93] Most senators realized that the primary purpose of the bill was to harm Cyrus Field, and they did not find his shifting his residence from the Bay State to New York reason enough to injure him. Nor did they buy the argument that the NYAP discriminated in favor of New York and against Boston.

It was in the power of the telegraph, explained a senator from Berkshire County, to make news flow rapidly and reach over the whole country simultaneously. "As far as I understand the system at present existing, it is wonderfully perfect. It operates with marvelous rapidity, and distributes the news to all localities at the same time . . . The news is simultaneously published in all these places; the merchants of one city do not get the advantage over the merchants of any other city." By establishing a competing line, argued the senator, "instead of one well considered and well adapted system, we shall have the system broken up . . . which will throw the whole thing into confusion."[94] Craig's early notion that the management of the nation's news required one unified system was obviously beginning to make sense to Americans.

The association's agent "*edits* the news of the world for this country," argued an attorney for the Boston merchants in support of the bill.[95] True, agreed an opposing senator, yet he thought there were worse options.

> I think the interest the press has in this question is paramount to all other interests in this country. It is to the press and by means of the press that the European news is communicated to every city of this country, and I think we should do nothing to place this intelligence in the hands of the merchants first, and afterwards in the hands of

> the press. I believe that the press of this country have a priority of right, for they will give it to the whole country simultaneously.[96]

The stand of the Boston press in the controversy demonstrated the depth of the transformation in the country's news-gathering system during the course of the decade. When in 1851 Craig first tried to bring the friendly association of Boston morning dailies to his plan of a national news service, they revolted and severed all their connections with the NYAP. After a short and costly contest, they had reconsidered, surrendered to Craig, and joined his scheme.[97] They realized that a single national organization for telegraphic news was indispensable. The position of the Boston press in the controversy of 1859 over the competing cable proved the strength of that conviction. With the exception of the *Boston Courier*, the papers stood squarely behind the NYAP. The *Boston Advertiser* noted maturely that the one renegade newspaper gained "no solid reputation for superior 'enterprise' by grumbling continually at *evils that are irremediable.*"[98]

The opposition of the Boston press to the Massachusetts cable bill was an important factor in its ultimate defeat. The bill was promoted as a means of placing the press and the public in Boston on a par with New York. The Boston press argued simply that the AP gave no particular information advantages to New York. "I do not understand the press of New York . . . to be the Associated Press," offered a senator from Berkshire County during the course of the debate. Rather, he believed that "the Associated Press are the combined press of every city of this Union—Boston, New York, Philadelphia, Baltimore, Washington, New Orleans, etc., etc."[99] Indeed, by 1859 the New York Associated Press had established a truly national system for the flow of public information.

6

"To Liberate the Newspaper Press": The New York Associated Press at War

The dominant position achieved by the Associated Press in the national news business invited challenges.[1] Contemporaries, who had very little exposure to organizations operating on a national scale, had difficulty perceiving and appreciating the real sources of its power. Its position on the fringes of the newspaper and telegraph industries appeared vulnerable to strong forces in each of those sectors. The AP was dependent, to an important extent, on the telegraph industry for collecting the confluence of American news and serving it to the press. Then too, the association was merely an adjunct to seven New York dailies, and neither its machinery nor its independent resources appeared overpowering. Each of the dailies that constituted the association had, by the early 1860's, more employees and larger budgets than the AP.

Indeed, telegraph leaders and member newspapers of the AP repeatedly tried to take over the association and integrate the functions of a wire service into their own operations. Yet the vulnerability of the association was illusive. By establishing a national system it had acquired a strategic position that could not be easily challenged. Only with the rise of parallel institutions of a national scope was the NYAP seriously threatened.

In 1859 and 1860, the NYAP and the telegraph industry were embattled in controversies reminiscent of the Craig-Smith melodrama of a decade earlier. As the nature of news and the character of telegraphic technology dictated, the battle lines were drawn once again over first possession of the news and a hold over the telegraph terminus, exclusive control over segments of telegraphic routes, and power in the consumer market for news. Courts, legislatures, and public opinion still played secondary roles, and the main theater of operations remained the control of foreign news.

Yet by the end of the 1850's some things had changed. Great telegraphic oligopolies replaced the small latifundia of bickering telegraph lords. The distribution of news was now a business of national scope, rather than a spirited competition between individualistic editors, news merchants, and small associations. The tension between monopoly and competition in news, therefore, expressed itself in new forms.

This time the instigator of hostilities was Daniel H. Craig. The Nova Scotia Telegraph Company was the one link between New York and the foreign news that was not controlled by his friends at the American Telegraph Company.[2] The line, sold by the government of Nova Scotia to private entrepreneurs in November 1853, was totally dependent on AP business. Upon acquiring the line, the new company picked up the government's contract of 1851 with the NYAP but, to remain in the black, found it necessary to double the rates for the association's foreign news dispatch (making them three times the full commercial rate).[3]

Spoiling for a fight, Craig arbitrarily stopped paying the line the guaranteed sum of $150 on the arrival of those steamers that his news yachts succeeded in intercepting at Cape Race.[4] The AP guarantee was for 3,000 words of fresh European news, explained Craig, not for news made stale by the costly operation off Cape Race.[5] If it was to survive, the Nova Scotia Company obviously had to seek other news customers to counteract Craig's move. Yet between the terminus of their line and Boston there were more than 500 miles of telegraph wire controlled exclusively by the American Telegraph Company; and Craig, they discovered, had "some strange power over that company."[6]

At this juncture, the interests of the Nova Scotia Company coincided with those of its former superintendent, Frederick N. Gisborne, and its bitter enemy of a decade earlier, Francis O. J. Smith.[7] The two men were promoting the Cape Ann-Yarmouth cable that would connect Massachusetts with Nova Scotia directly. Smith and Gisborne's cable project promised to free the Nova Scotia company from dependence on the American Telegraph Company. With the cable project gaining support from important interests in the Bay State, the Nova Scotia company felt secure enough by 1 November 1858 to serve the NYAP with six months' notice prior to termination, as stipulated by their contract.[8]

At this time, Michael A. Zabriskie and George W. L. Johnson—together, the United States AP—entered the competition. Johnson was one of the earliest telegraphic news vendors, and by 1851 he was a prominent news supplier to the press between New York and Washington. Craig's national system of news gathering threatened to crowd him out of the field. In 1850 he appealed desperately but unsuccessfully to the Nova Scotia authorities for the same rights as Craig in procuring steamer news. He also arranged with a Pittsburgh telegraph operator to steal AP news reports from the South. This "news conspiracy" was, however, exposed prematurely and Johnson had to abandon the fight.[9] Less is known of Zabriskie, a young man from New Jersey, probably a newcomer to the industry.[10] To Craig, Zabriskie appeared to have "more money than brains." As it turned out, he had significantly less of the former by the time his dealings with Fog Smith and the Nova Scotia Company were over.[11]

Picking up the AP's contract with the Nova Scotia telegraph on the same financial terms, Johnson and Zabriskie were granted some additional privileges. Whereas the AP was allowed uninterrupted transmission of 3,000 words of steamer news only after presenting its dispatch, the new firm was given priority over all foreign news dispatches from the moment the steamer was sighted. Moreover, the telegraph company agreed not to send any competing reports for some twelve hours following the transmission of Johnson and Zabriskie's dispatch.

Just as Johnson and Zabriskie were making their arrangements with the Nova Scotia Company, the negotiations between the American and Morse and Kendall's Magnetic for a consolidation of their systems, which had seemed all but successfully concluded, reached a crisis. The ex-president and director of Magnetic, William M. Swain, editor of the prominent *Philadelphia Public Ledger*, tried to win over the press between New York and Washington from the NYAP, offering seceding newspapers dramatically lower rates than those it charged the customers of the NYAP. This offer, combined with the prospects of receiving early foreign news through Zabriskie and Johnson, made an attractive package. Virtually all Philadelphia's papers entered the new arrangement, knowing full well that by doing so they stood to lose their privileges as clients of the New York association.[12]

The AP's constant fear that competition in telegraphy would lead to competition in news gathering now appeared fully warranted. With the termination of the AP's contract with the Nova Scotia Company drawing near, a summary review revealed a mirror image of the telegraph-press alignment of the contest of 1849. It was now Craig who appealed to the Nova Scotia telegraph on the merit of the first come first served principle, and the small rival association that held priority over the wire-head. It was now Craig who had "superior facilities" over connecting lines; and it was Smith who was trying to arrange for competing facilities to thwart the enemy's monopoly over many miles of wire "between us and the news." There were, however, several constants.

The AP's sway over public opinion was, if anything, greater than in 1849, and Craig was ever-capable of "superior activity."

Where O'Rielly had succeeded in 1849 in creating an opposition line, Smith failed in 1859 to establish an alternative connection between Halifax and Boston. The press and the public in Boston were opposed to the Cape Ann-Yarmouth telegraph project. After an acrimonious debate in the Massachusetts legislature sharply focused on personalities, that body declined to grant a charter to Gisborne and Smith's cable project.[13] Yet even after this failure, Smith and Gisborne and Kendall and Swain still considered their position strong. On 6 May 1859, the first of the fortnightly European mail steamers arrived at the port of Halifax. The foreign news dispatch of the NYAP was first in the office but was denied transmission. By the time Zabriskie and Johnson filed their own dispatch, an hour later, a fast horse express was carrying Craig's message across Nova Scotia to Sackville, the northern terminus of the New Brunswick line, controlled by the American Telegraph Company.

Craig had forgotten none of his old tricks. At Sackville, where the American Telegraph Company took over, it was found impossible to transmit Johnson and Zabriskie's dispatch to New York. One wire was conveniently out of order, and the other was engaged in transmitting to the NYAP the full details of what happened "In the beginning" and thereafter, until the express from Halifax arrived.[14] Craig's dispatches, fourteen hours on the road, were the first foreign advices to be published west of Sackville, New Brunswick. On subsequent arrivals, the American Telegraph Company simply declared outright it would not transmit Johnson and Zabriskie's message ahead of Craig's.[15]

Sharp complaints were voiced in telegraphic circles against the American's violation of the first come first served principle. The company's main defense was that it operated under a New York charter, and that state stipulated priority to newspaper editors and publishers; strictly speaking, Zabriskie and Johnson were not editors but private entrepreneurs.[16] The press of the country, north, south, and west of Philadelphia, was generally supportive of the American's policy and critical of the Nova Scotia line. Not that it had much of a choice; Craig had the foreign news first, he was the best source for domestic news, and he did not tolerate criticism very well. Yet morally, too, Craig appeared to have won the contest. As in 1849, he beat his rivals to the news and was first in getting it into the telegraph office. He was deemed justified in having it reach New York first, contractual privileges notwithstanding.[17]

Craig celebrated his victory by banning the *Philadelphia Public Ledger* from the NYAP's news service. The expulsion symbolized the victory of the AP's American Telegraph Company over Swain's Magnetic and Fog Smith's Union line.[18] Yet Craig's victory proved to be short-lived.

Beginning in October 1859, the NYAP no longer had to worry about rivals of the American Telegraph Company supporting a wire service in competition with its own. A series of agreements entered into that month effected a complete consolidation of the telegraph facilities along the seaboard, from Sackville to New Orleans. Craig, after a taste of the destructive potential of telegraphic competition for a national news service, strongly supported the consolidation.[19] It represented the realization of his vision of the natural evolution of the telegraph industry toward monopoly.

Yet the enemies of the AP remained active. The Magnetic Telegraph Company entered the consolidation still angered by Craig's high-handed treatment of their director, William Swain, and his *Public Ledger*. Nor did they entirely forget the other little tricks Craig had played on them during the course of the decade, as rivals of his telegraph interests. The Magnetic therefore made the consolidation conditional on strict adherence to the first come first served principle and on an effort to "afford increased facilities" for news transmission.[20] Article 11 of the contract called on the consolidated company "to use all legitimate means *to prevent monopolies* of market and other news entering the United States from foreign countries, putting up, if available for that purpose, additional wires upon their lines, or constructing an additional line or lines . . . leading to Halifax and other landing places of ocean steamers which may be used contemporaneously *for the accommodation of rival interests*."[21] The agreement even called for the revival of the Cape Ann-Yarmouth cable project. At the time, no one understood these articles to be more than a face-saving ploy. Yet Kendall privately stated that ending his life in a "determined effort to break up this infamous combination" would be for a worthy cause. The association's power "to crush every independent telegraph line and reduce the Press to a condition worse than that of France" made Kendall's "blood boil."[22]

The other major addition to the consolidated company was a group of telegraph entrepreneurs led by Robert W. Russell, an Englishman. They owned the House line between New York and Washington, as well as various telegraph lines in the South, and the rights to the House patent nationwide. Russell too had on previous occasions experienced the questionable pleasure of having differences with Craig.[23] He had no special sympathy for the man who had publicized him throughout the country as a "soulless vampire," a "loathsome reptile," a "slimy vermin," and a host of other imaginative epithets, some more deserved than others.

Gradually, Russell became the "head and limbs" of the American.[24] With the consolidation satisfactorily accomplished, Field, Cooper, and Hewitt preferred to shift their attention to the cable project and to their other, perhaps less exasperating, business interests. They sold much of their stock in the American to Russell. With Russell in control, Craig's creative efforts to support a telegraph monopoly that would act in concert with the news monopoly appeared to have backfired.[25] The American under Russell's leadership decided it had "one course to pursue and that is to relieve itself from the incubus of the New York Association and its General Agent,"[26] and it pledged to "liberate 'the newspaper press' from the yoke of D. H. Craig."[27] The monster had risen against its creator.

"Able, imperious, willful and persevering," Russell had a vision of his own concerning the proper relation between the wire service and the telegraph. In leading the fight against the NYAP he apparently was not seeking personal revenge on Craig. Russell was trying to map a perfectly rational course for a telegraph monopoly to pursue regarding the supply of telegraphic news. His stated policy that "the NYAP must be broken up" was backed by a plan that prescribed the "means . . . to accomplish that result."[28]

Unlike the leaders of the Magnetic, who thought expanding telegraph facilities would make competition possible, Russell understood well that there was more to the association's power than control over bottlenecks in the transmission of information.[29] It was virtually impossible to compete with the economies of scale that the AP achieved through its national system. He therefore decided to beat the AP with large increases in telegraphic tolls. Increased tolls would weaken the association while expanding his company's treasury. Holding a well-entrenched monopoly over seaboard telegraphy, Russell believed that there was no danger that the price hikes would encourage competition. Furthermore, Russell wisely accompanied the rises with a change in the price structure devised to diminish the economies of scale enjoyed by the NYAP.

With the standing price structure, each client newspaper paid a relatively small sum for receiving the consolidated report, whereas the association was charged high tolls for transmitting the crude news it gathered into the New York headquarters. This made competition difficult; an upstart rival agency would have to make a heavy preliminary investment in news gathering before it had a chance to attract a sufficient number of clients to cover its costs. Moreover, a newcomer would find it difficult to attract customers by reducing their already minimal charges.[30]

Russell decided to turn this price structure on its head. By making it cheap to procure news and costly to receive the AP's consolidated report, he believed "other associations would be formed, in and out of New York."[31] Better yet, newspapers would compete in procuring special reports. Russell anticipated that "numerous versions of the news would be sent by different reporters; some of the papers would vie with each other in the fullness and completeness of their reports, and in the style of the same." By breaking this monopoly of knowledge the American would perform a great service to the public but, as Russell assured his stockholders, it would also have some other salutary effects: "Our lines in Maine, New Brunswick, and Nova Scotia would cease to be pauper lines, starving upon a single message sent by Mr. Craig once a week or once a fortnight, for a miserable compensation, and that accompanied, too, by gross and brutal insults."[32]

An enthusiastic Fog Smith, now a director of the American, victoriously explained to whoever would listen that that had been his grand design all along. Competition between numerous news-gatherers was an important step toward what he viewed as "perfection in the relations of the telegraph and the public," namely, having "each member of the community" pay the telegraph company for his news, rather than paying two cents to an editor and reading the telegraphic column in the newspaper.[33]

Backed by the executive committee and a majority of the board of directors, Russell discharged his first salvo at the AP, aiming at its loyal Boston affiliate. Prices for receiving the NYAP report by the Boston Association and the way papers between Boston and New York were raised by as much as 150 percent.[34] Craig pointed out in vain that the AP was paying three cents a word, even more than the full commercial rates, for transmitting news in the opposite direction, from Boston to AP headquarters in New York.[35]

Russell, however, acknowledged that an important source of the AP's power and the key to the resolution of the whole issue was control over the foreign news. He declared he had a secret "programme" for breaking the association's lock on the European news, which he was not as yet willing to disclose.[36] He did secure a lease of the Nova Scotia line but did nothing to offer greater facilities to make competition in foreign news possible. He strung no additional wires on the Nova Scotia line, nor did he show any interest in leasing the Newfoundland line. Yet Russell vowed repeatedly to take active measures to break up the AP's monopoly on the foreign news.[37]

Russell's program was perfectly patent to all active players in the telegraphic news arena: it was for the American Telegraph Company "to get up another system of reporting, both at home and abroad."[38] "They think," the *New York Times* explained to its readers, that "*they* should be allowed to furnish the news to the public instead of the newspapers."[39]

Craig responded with an ultimatum: he demanded "the ignominious discharge of Russell" and his allies "from the positions they disgrace in the management of the Company."[40] If this didn't happen, the AP would build its own trunk line between Boston and Washington. This time, Craig appeared to be fully backed by his superiors. Once the proprietors of the AP realized that Russell's threats were in earnest, they showed their alarm by demanding to be represented by five editors on the American's twelve-man board of directors. The emergency made them acknowledge at last the extent of the interdependence between their news messages and the medium for their rapid transmission.[41]

Craig thought that electing five editors, who were sure to be "intelligent" and "respectable," was a good idea. Granted, they lacked experience in telegraphic matters, but still they would "contrast favorably" with the "Lottery Swindlers, Pettifoggers, Illiterate Cheats, Rum Sellers, Drunken Sots, and Superannuated Political Knaves" who had sat on the board.[42] When the demand of the editors for representation was scornfully turned down, they went one step further and supported Craig's idea of building an opposition line. Realizing how easy it was to trespass on news gathering from the realm of telegraphic transmission, they were now willing to venture across the border themselves and move from newspaper ownership to telegraph ownership.

Even the chorus of approbation from the outside press that followed Craig's ultimatum this time appeared genuine. Craig may have been a "hard task-master" for the press of the country, but he was also a representative of the press. Editors realized that once the telegraph tried to maximize its revenues and entrench its monopoly by exercising control over telegraphic news, they would not be any better off.

Russell was correct in pointing out to the nation's newspaper editors that the AP, with its rule that forced clients to receive its reports exclusively, confined the entire press of the country to one version of the news. That version, suggested Russell,

> may be bogus in many important particulars, but no correction can be had through the public prints. Private messages, sent to particular individuals, may correct the errors in the published dispatch, or may furnish additional details, but the public at large must put up with what Mr. Craig gives them. No contradiction—no explanation—no further details can be had from other sources by telegraph.[43]

Kendall was even blunter than Russell in his warnings to the press of the evils inherent in the AP's monopolistic control over the national flow of news. Craig, he claimed, "makes use of these dependent presses to circulate, under cover of news dispatches, any falsehood which may tend to accomplish his designs or gratify his malignity." The purpose of the American Telegraph Company was not only to decrease the association's income but also to destroy its "power over the American Press out of New York . . . and [cut] off his [Craig's] means of misleading the public by false dispatches."[44]

Yet Kendall should have realized that his argument was a double-edged sword. It applied to news control by Russell no less than by Craig, and between the two evils, the press had no doubt where its best interests lay. All the examples Kendall offered of Craig's filling the press with falsehoods had to do with struggles between the AP and telegraph companies. To the extent that Craig used his monopoly of information to control the contents of his reports, he did so in the interest of an association of respectable newspapers.

The problem of one news report's satisfying divergent interests was genuine, but there were certain checks on the reliability of Craig's reports and certain balances on their possible bias. Craig himself had pointed out, in response to similar charges, that

> the affairs of the Associated Press are managed under the immediate direction of the editors and proprietors of the seven leading morning journals of the city of New York—editors whose well known *diversity of opinion* upon every prominent subject which agitates the public mind [*sic*]—proprietors and editors whose shrewdness and rivalries are the best guaranties the public can have for their watchfulness of the acts of their general news agent.[45]

The press of the country preferred to remain under the yoke of the seven wise editors of New York than to have telegraph barons hand down the "one version" of national and international news. They expressed this preference in a barrage of letters, articles, and resolutions of regional conventions of editors.

There was no doubt in telegraphic circles that Craig was dead serious in his demand that the American Telegraph Company oust its leader. The real question was whether the threat to build competing telegraph lines was a viable option or a bluff. The managers of such a project faced the difficulty of finding a telegraph patent not owned by the American to employ on the projected line, and a convenient but unoccupied right of way along which to stretch it.[46] Further doubts involved the editors' ability to assume the large financial responsibility involved, or to find investors who would risk their capital on the project.

Russell tried to relieve his worried stockholders by asserting that such a line "would not pay half its expenses, with all the patronage D. H. Craig could bestow on it."[47] The stockholders, assured by Russell that "the Associated Press business is not worth having," must have blinked when they took a close look at the relevant figures. The AP business between Boston and Washington amounted to some $100,000–$150,000 a year.[48] If $55 was the annual expense of operating each mile of wire, an AP two-wire telegraph line between Boston and Washington would require only some $55,000 annually.[49] Even after paying a fabulous dividend to the lucky investors and before sending a single commercial message, Craig's estimate that the press would save $50,000 annually on telegraph tolls was conservative. A Boston entrepreneur guaranteed that if the AP notified him in the morning that it wanted a line, "the stock would be subscribed for in Boston before two o'clock that afternoon." Considering the impressive numbers Craig posted for his business, such a guarantee was hardly extravagant.[50]

Although the American's executive committee dismissed the patronage of the press as yielding "not much more than a single second rate town in New England," adding up the figures showed that the AP paid more than $250,000 in tolls to the company.[51] The total annual receipts of the American (extrapolated from the results of the first half of 1860) were $970,790.[52] The press thus yielded between a quarter and a third of the total receipts of the company. In other words, the AP business meant the difference between bountiful dividends and a net loss.[53]

The crucial issue concerning the feasibility of a new line was the problem of establishing connections with other telegraph companies. To ensure themselves jointly against the invasion of wildcat telegraph companies once the Morse patent expired in 1861, the major telegraph companies in America combined to form a defensive alliance, the North American Telegraph Association. Their six-party contract (better known as the "treaty of the six nations") guaranteed exclusive connections between the parties to the agreement, thus making hopeless the prospects of establishing an isolated line to compete with one of the members.[54] But an AP line would hold an entirely different position from any newcomer the treaty had anticipated. It would have a large business committed to it from the start and would enjoy solid public support. Should a press line materialize, it was doubtful if any of the interior lines that were party to the treaty would decline connecting with it out of loyalty to an agreement Russell himself had already alluded to on several occasions as "a rope of sand."

Whether Craig was bluffing or not was therefore to determine the fate of the six-nations' treaty. The general opinion in the telegraph industry was that Craig was serious, and that conviction sealed Russell's fate. The five western "nations," led by the Western Union, began putting pressure on the American to set in order its relations with the press before the AP's projected line destroyed the status quo. "The whole telegraph interests of the country are put in jeopardy by that miserable stubborn man Russell," wrote Western Union leader Hiram Sibley to a colleague.[55] The Western Union left no doubts that its sympathies were with the press and that it valued a connection with a line operated by Craig more than a connection with Russell's network of wires. The stockholders of the American well understood that such a coalition would destroy their company.[56]

With the pressures on the American from without mounting, the internal opposition group led by Field and Hunt decided that their investment in the American was no longer safe. They therefore liquidated their remaining interest in the company that they had founded together with Craig. Russell, determined to fight to the end, tried to fortify his faltering position by purchasing their entire stock at an advanced price. He filled the vacancies on the board of directors occasioned by the departure of the New York group with two strong and purposeful men, Cambridge Livingston and Colonel Edward S. Sanford, manager of the Adam's express. A showdown was imminent, and it was Craig's turn to show his hand.[57]

When he did, the telegraph industry not only gasped with surprise but probably also sighed with relief. In the clamor of the contest between the American Telegraph Company and the press, onlookers seemed to forget the real power of Field's position. The American's hopes for future success were dependent on a future Atlantic cable, and the cable was Field's project. Besides, "too big to touch" and very far away, the Newfoundland line, capitalized at almost as much as the American Telegraph Company, had all along been under the exclusive control of Cyrus Field's group. On 5 October 1860, the New York dailies quietly announced that the New York, Newfoundland, and London Telegraph Company was leased for five years, or until a successful cable was laid, for an undisclosed sum. The lessee was identified as the NYAP.[58] Russell's ambitious scheme to supplant the NYAP in the supply of European news to the press, and ultimately, of all telegraphic news, was now history.

With the increase in steamer traffic to New York during the 1850's, the news brought by the fortnightly Cunarders touching at Halifax constituted a declining proportion of the trans-Atlantic news flow. By 1860, facilities for intercepting steamers off Cape Race had been greatly improved, and steamers bound for New York, as well as Halifax, were game for the association's Newfoundland news yachts.[59] The *New York Herald* explained that the press, by virtue of its "new and extensive arrangements with the Newfoundland Telegraph Company," would be in a position "to obtain intelligence from off Cape Race from every steamer outward and inward bound that passes that point."[60] The *New York Times* similarly concluded that "hereafter, nearly all the European news will be landed at Cape Race"; and it added, sarcastically, that "the American Telegraph Company cannot *just yet* go into the business of speculating in foreign news. They will be under the necessity of confining themselves for a short time longer to their legitimate business, of forwarding messages, at a fixed price, and in regular order, for persons who choose to send them."[61]

By the end of 1860 the AP again dominated early receipt of the foreign news, and it once more commanded the wire terminus. So much for Russell's brash threat to wrest the foreign news from Craig's hands and, by so doing, liberate the press of the country. With its failure to supplant the NYAP, the American had no alternative but to retreat. It had to mend its relations with the Western Union and its other partners to the treaty of the six nations, as well as with the press. Craig's outrageous demand that the American Telegraph Company oust Russell if it wished to resume its good relations with the press was complied with. Even Edward Sanford and Cambridge Livingston, who occupied board seats on the power of Russell's stock, realized that the good of the company required them to join the majority that elected a new executive committee. Robert W. Russell was not among the executives that shortly concluded a comprehensive contract with the NYAP.[62]

Yet the larger issue of the ultimate relationship between the press and the telegraph remained unsettled. The downfall of Russell proved that the press and public opinion were too strongly opposed to permit the telegraph to control the news. Craig had obviously failed in his attempt to form a telegraph monopoly that could be counted on to act in the interest of the press association. The Civil War postponed the resolution of the tension between the telegraph and the press for a number of years. Not until that dispute was settled did the conflict between the telegraph and the press reach a permanent, if uneasy, resolution.

President-elect Abraham Lincoln departed Springfield on 11 February 1861 after an emotional farewell address in Great Western Station. He was on his way to Washington, and Americans were curious to learn of his thoughts just before he assumed his new job. Aboard the special train, Lincoln handed the notes of his Springfield farewell to Henry Villard, a journalist with whom he had been acquainted since his senatorial campaign against Douglas. At the first stop, Villard telegraphed the text of the address to the AP in New York; from there it was dispersed throughout the country. It was thus through an AP dispatch that a tense country got an authorized version of its new leader's long-awaited words.[63] Four years later, a shocked nation learned from an AP report that "the President was shot in a theater tonight, and perhaps mortally wounded." The message was telegraphed from Washington by the association's veteran and sometimes erratic Washington agent, Lawrence Gobright, who beat the host of energetic Washington reporters to the final story of Lincoln's administration.[64]

These two episodes, spanning Abraham Lincoln's presidency, were in one respect representative of the performance of the AP in the Civil War years. The wire service was an authoritative source for news throughout the military conflict. The government supplied it exclusively with its bulletins and official announcements, and the AP assumed the position of a semi-official medium for the distribution of information emanating from sources in the administration.

Yet, in a different sense, the two episodes were entirely uncharacteristic of the journalistic role of the AP in the conflict. Rarely during the war did the association record news scoops. Generally it was not the AP but individual newspapers that broke the primary war news to the public. Indeed, the war liberated the press in a large measure from its complete dependence on the AP for its fast news.

Before the war, the *New York Herald* employed four reporters in Washington and two in Albany, but along with them only three or four "roving correspondents." These correspondents, explained the *Herald* reporter William Shanks, were expected to "go from point to point on the scent of news," and to "'be in at the death' wherever news is to be had." But as a rule, they were to be found only "where the AP agents are not." During the Civil War, the *Herald* reinforced its corps of roving correspondents; it ultimately numbered sixty-three "specials" seeking military news.[65] Obviously, there were a great many news-breaking locations where the AP was not represented.

The official history of the AP proudly stated that the organization had more war correspondents than even the biggest newspaper. If the *Herald*, with its sixty-three reporters in the field, is given that distinction, the AP history appears to have exaggerated the real number of its wartime reportorial staff about tenfold. In his annual report for 1862, Craig mentioned a regular agent in Fortress Monroe and three agents with the army. He may have employed one or two additional reporters with the Army of the Potomac earlier in 1862. Various lists and references to AP agents make it clear that this small representation in the field was not expanded much in the latter years of the war. Later descriptions of Craig "keeping his army of correspondents on the move from one war theater to another" appear to reflect not much more than AP bravado.[66]

The meager digest of war news served up by Craig's diminutive "army of correspondents" did not satisfy the press's demand and encouraged competitors to enter the field. Three young military reporters believed it was in their power to supplant Craig's service. Horace White, Adam S. Hill, and Henry Villard formed what Villard considered "the first news agency in competition with the AP." Although they were "bitterly attacked by the AP for disturbing its monopoly," they succeeded in attracting many clients.[67]

Whatever the AP did in the way of news gathering in the field resulted from its obligations to provide the outside press with news. Its own New York members were spending enough money gathering their own war news. They did not wish their weekly bills from the AP to reflect any additional large expenditures in that field.[68]

The balance sheet of the association indicates that business continued much as usual during the first full year of hostilities—so much so that operating expenses actually decreased. During 1862, a year in which each of the seven member newspapers was spending between $60,000 and $100,000 on war correspondence alone, the entire expenses of the AP were about $124,000, of which some $20,000 was spent on news from Cape Race.[69]

The consumer market for war news revived individual enterprise and competition between dailies at the expense of the cooperative arrangements. Before the shelling of Fort Sumter, news about "trade, commerce and business, which concern every man in the community," was what sold newspapers. The AP was the prime source for such news and an important supplier of information about legislative maneuvers, political intrigues, and compelling items of general interest.[70] Once the war broke out, such news was preempted by news of the war, which affected everyone in a much more immediate and decisive fashion. "The grand object of the true journalist, is to be fullest, promptest, and most correct on the one uppermost topic of the hour," observed James Parton. "The paper that does this oftenest," he added, "is the paper that will find most purchasers." During the Civil War, "this fundamental truth of journalism was apparent to every mind."[71] Circulation, as a rule, increased "in exact proportion to the interest of the news." When an important battle was taking place, newspapers could expect to sell five times as many copies as on other days.[72]

Dailies could not be expected to pass up the opportunities to increase their circulations dramatically by beating their rivals to the news. The stakes were too high to allow strict obedience to the AP rule requiring members to make their specials available to their neighbors. It was also contrary to every sense of justice that news procured with great effort and sacrifice should be shared merely because it was transmitted by telegraph. Nor was the enormous price of gathering such news shared fairly, since the other members paid only their portion of the relatively small cost of telegraphing the news to New York.[73]

Far more important, the AP could not offer an immediate alternative to individual efforts by its members in the quest for news from the battlefields. The nature and structure of Craig's particular news-gathering system were not suitable for covering a war. The association's static "news net" paralleled the network of the country's telegraph lines.[74] When armies marched in the field, they passed the outposts of the commercial telegraph system and were beyond the reach of Craig's network of stationary agents. Furthermore, in Craig's philosophy of operating a wire service, his men were not expected to gather primary news; they were agents, not reporters, and their function was to collate news gathered by the newspapers of the country and transmit it to New York. Craig did station agents at a few strategic points (such as Liverpool, Cape Race, and Washington) to gather primary "routine news" directly. Antietam, Chickamauga, Gettysburg, or Appomattox was not among them.[75]

Yet the AP as a cooperative system managed to survive under these pressures. Even in the radically altered circumstances of war journalism, the advantages of the cooperative arrangement to each member were many. The members of the AP ultimately found ways to harmonize their individual interests with the advantages of their common association. They succeeded in restoring competition without even formally breaking the association's peacetime rules.

For military operations in the East, Washington was the first telegraph facility available for transmitting the news from the armies to New York. The rules of the association permitted newspapers to transmit special reports from Albany and Washington without having to share them with the other members. This was clearly intended to accommodate the differing political loyalties of the member newspapers in reporting political news; but with the outbreak of hostilities, it permitted the Washington bureaus of the several dailies to transmit much of the war news to New York exclusively, without breaking AP rules.[76]

Although telegraph censorship was an obvious drawback for war journalism, it also facilitated the transition from cooperation to competition in news gathering. Censorship in the North throughout the Civil War focused on telegraphic messages. In many cases correspondents returning from the field with dramatic news, or picking up sensitive scraps of information in Washington, did not even try to telegraph them to their editorial headquarters. Instead, reporters exhibited wonderful enterprise in trying to personally deliver their news to their home offices. The main incentive for expending fortunes on transporting news was of course to escape the "scissors of various sizes," the "full rack of pens, several inkstands . . . lead pencils of different colors—black and blue—[the] pot of mucilage," and the "other properties" of the censor.[77] But in evading telegraph censorship reporters also evaded the rule compelling members of the AP to share news messages received by telegraph.[78]

Although the leadership of the AP in news gathering was weakened by competition, the war made the AP the foremost channel through which the administration addressed the nation. Washington recognized the potential usefulness of the AP—the first mass communication medium of national scope—as an indispensable tool to access a national constituency simultaneously. Consequently, the AP became a cornerstone of the Lincoln administration's strategy of information management.

Thus, in 1864, the violently anti-administration *Journal of Commerce* complained bitterly that since the beginning of the war it had published more than $50,000 worth of what it considered "gratuitous advertising" for the administration. "There is scarcely a night in the year," stated the *Journal*, "that the War Department does not send us, on manifold paper . . . [some] proclamation or order."[79] The reference was of course to AP matter. The *Journal*, however, was giving only one side of the story; the other was that government war bulletins and releases were generally of very great news value. Sometimes they included the first information of important military events blocked for days by the censor, or they announced dramatic political or administrative developments. "As false as a bulletin" became proverbial during the war, but given the meagerness of alternative sources of information, a monopoly of the bulletins was a tremendous advantage for the AP.[80]

The give-and-take between the association and the administration was not limited to official bulletins. This became apparent from the earliest days of Lincoln's beleaguered administration. During the Sumter crisis, the authorities were in a dilemma. On the one hand, it was important to keep the public informed in order to retain its confidence and support in the critical policy decisions being made. On the other hand, the president, trying to "thread his way between war and disunion," his cabinet divided, could hardly afford to commit himself to a rigidly formulated plan of action in the fluid state of events.[81]

The issue came to a head early in April. "Wanted—A Policy!" proclaimed the *New York Times* at the head of its editorial of 3 April 1861, and opinion throughout the North called for a firm response to Confederate demands to surrender Fort Sumter, where supplies were running low. On 9 April the president finally ordered an expedition to Sumter that was capable of either provisioning or reinforcing the fort. Although he wished to inform the Union of the fateful decision, Lincoln preferred to convey his intention to Confederate authorities confidentially, rather than through a public declaration.[82]

AP Washington agent Lawrence Gobright resolved the dilemma for the administration. Well informed about the state of affairs through his daily contact with the nation's leaders, Gobright wrote a comprehensive statement of the administration's action as he understood it. He sealed these "notes for a telegram [to the AP]" in an envelope, wrote on it "important—not an Application for Office," and handed it to a cabinet member whom he requested either to destroy the notes or to confirm their contents. The same day, Gobright received official endorsement of his dispatch and it was circulated throughout the country. The statement, released only three days before hostilities began, was as Gobright noted "the first semi-official announcement of 'the policy of the administration.'" It was indeed in the form of a semi-official organ that the relation of the AP to the government finally crystallized.[83]

The flow of information, however, was not one-way, from an administration wishing to reach a large public, to the AP, craving for news.[84] Throughout his administration, Lincoln gave the distinct impression of a man on a quest for information. His close and informal relations with the Washington press corps and the government's censorship of the telegraph brought into his possession much information that the press had acquired. Gobright's memoirs indicate that Lincoln, "seeking after information," was his regular partner in the information chase. Although Gobright and other reporters probably exaggerated their intimacy with the president, after the assassination had "translated Lincoln into sainthood," it appears that the president and the agent had exchanged news bits on a regular basis.[85]

The wire service and government nexus was managed through the most simple machinery, and it focused on Lawrence Gobright's daily routine. Samuel Wilkeson, head of the *Tribune*'s Washington bureau, described this routine and its implications to a House investigative committee early in 1862. Gobright, testified Wilkeson, with explicit factuality and implicit envy,

> goes to the several departments and inquires for news and they give him all orders which have been issued suitable for publication, and in as much as his dispatches are sent north west and east and indeed all over the country where this Associated Press arrangement exists, the government whenever it wants to make an impression on all points within twenty four hours makes him the vehicle of intelligence.[86]

Like most other Americans, the committee members were apparently unaware of this arrangement. Yet, as their further questions to Wilkeson indicated, they fully grasped its significance:

Q. He [Gobright] is then, to a certain extent, an agent of the government?
A. He has become . . . pretty much of a government institution. They make use of him for the public diffusion of intelligence of the war and the policy of the administration. His functions have changed within six months. He has become almost as much the agent of the government as the agent of the newspapers.
Q. Do the Departments intrust him with information which they decline to give or withhold from others?
A. That is a thing of almost daily occurrence.
. . .
Q. Has he access to the Secretaries?
A. Yes Sir, he has access to every department.
Q. And to the President?
A. Yes Sir.

Through subsequent questioning the committee members learned that the administration could and did use the wire service to disseminate false information, so as "to produce an impression" favorable to the government "all over the United States."[87]

The semi-official relationship between the association and the government was based on their mutual interest in information control. The government not only required a convenient channel through which it could communicate to the public, but it was interested in preventing the circulation of unauthorized information. The AP, for its part, was always comfortable in operating in the news market as a monopoly. Therefore the two parties had a mutual goal.[88] Washington authorities did everything within their considerable power to fortify the position of their semi-official mouthpiece in the national market for news: AP reporters had free access to the president and the War Department;[89] and more important, the association was given preferential access to telegraphic facilities.

In April 1861, the government assumed exclusive control of the lines of the American Telegraph Company merging in Washington, D.C. On 25 February of the following year, Secretary of War Edwin M. Stanton took military possession of all telegraph lines in the country.[90] The administration realized that the telegraph was the most powerful means of communication available, and the military censorship established under Lincoln's administration did not concern itself with anything other than telegraphic messages.[91]

Through telegraphic censorship, the authorities facilitated the dominance of the AP as news purveyor to the country. Early in the war, after short stints with the Treasury then the War Department, censorship responsibility came to rest with the State Department. Frederick W. Seward, son of, and assistant to, the secretary of state, instructed the censor "to prohibit all telegraphic dispatches from Washington, intended for publication, which relate to the civil or military operations of the government, with the exception of the despatches of the regular agent of the associated press."[92] These dispatches, with their blanket clearance, were the standard the censor was instructed to use for approving specials by other correspondents.[93]

In 1862 the congressional investigating committee inquiring "if a telegraphic censorship of the press has been established in this city" questioned Lawrence Gobright as to the privileges the association received from the administration for transmitting its news. To the extent to which he was willing to admit preferential treatment, Gobright attributed it to the even-handed nature of the association's reports. "My business is merely to communicate facts," he explained.

> My instructions do not allow me to make any comments upon the facts which I communicate. My despatches are sent to papers of all manner of politics, and the editors say they are able to make their own comments upon the facts which are sent to them. I therefore confine myself to what I consider legitimate news. I do not act as a politician belonging to any school, but try to be truthful and impartial. My despatches are merely dry matter of detail.[94]

This self-censoring association, by nature cautious and mainstream in its orientation, was an ideal partner for the government in news control through censorship.[95]

There were, however, other significant advantages for the government in using the AP as its disseminator of telegraphic news. If Gobright emphasized his innocence of sins of commission—his declining to add comment that would color the reports—others pointed out that the AP's notorious distaste for omission was just as conducive to the interests of the administration. As Sam Wilkeson testified, he himself would not permit any department of the government to dictate a dispatch to the *Tribune.* Gobright, in contrast, "will take and use what they give him." Indeed, he could be trusted to take even the most "oppressive" dispatches and "send them on without asking questions." He operated mechanically, failing to "exercise his judgement at all in regard to the interest and value" of the government-generated dispatches he passed on to the country's newspapers.[96] In this respect, what set Gobright apart from the two dozen other Washington reporters was, in essence, the difference in professional standard between simply a gatherer of news, responsible only to his sources and to his newspaper customers, and a newspaper journalist who addressed himself to the public.

Moreover, a technical aspect of the AP routine of news transmission effected a real advantage to sources bent on information control. The consolidated news dispatch emanating from AP headquarters in New York reached the country's newspapers late at night. During those hours the telegraph company's lines were virtually free of commercial message traffic, and comprehensive circuits could be put to the exclusive use of news transmission, at low cost. Late-night transmission was not only cheap, but it also ensured that the news reached newspapers just in time for hurried compositors to set up the dispatches, as transmitted, in type. This was long after opinionated editors, capable if not prone to tampering with official reports, had retired to their favorite saloon or bed for the night. With government dispatches published "before their worth [could] be scrutinized" and with no time for editors to examine them, government opinion got into type just as its spokesmen formulated it, unaltered and uncensored by the press. In this way, the AP channel ensured the administration of having the first chance to make a first impression on the public mind.[97]

Thus, wire service function and structure, policy and practice, were uniquely fitted to the publicity requirements of the government of a free nation at war. Collaboration with the AP provided the government with a channel for publishing its statements "in all the principal newspapers" instantaneously and simultaneously. Because the AP was deemed by the government to "represent the newspaper press of the country," the advantages conferred on it appeared to be legitimate.[98] The AP features of night service, the mild journalistic temperament characteristic of its uncritical employees, and the professional standards of self-censorship it displayed represented added bonuses to the government in its federation with the AP.[99]

The administration's effort to control the nation's channels of information was perhaps most dramatically exposed in the well-known episode of the "bogus proclamation" of May 1864.[100] Hoping to make a lot of money fast, Joseph Howard, a once prominent New York journalist well versed in Lincolnian prose and well acquainted with AP procedures, composed what purported to be a presidential proclamation following recent Union military debacles. The bogus proclamation recommended a day of fasting and solemn prayer and called for 400,000 additional troops. It was prepared and delivered by an accomplice to the offices of the AP newspapers in New York, on association forms and in full conformance with its procedures. Howard expected the report to cause a sharp rise in the price of gold, a good quantity of which he had purchased the previous day.

The fabrication was no more than a partial success. Only two of the AP newspapers, The *Journal of Commerce* and the *World*, published it on the morning of 18 May; these happened to be the members most vehemently opposed to the administration. The *Herald* was fooled into printing 20,000 papers with the proclamation but got back most of them before they reached the readers.

Early on the day of publication, all the facts concerning the deception, save the identity of its perpetrators, became known.[101] Craig immediately announced a $1,000 reward for information leading to the arrest of the parties who took the name of the AP in vain.[102] General John A. Dix, commander of the military department of the East, immediately began an investigation of the affair.

After all the details were revealed, showing that the *World* and *Journal* were the victims of a forgery, General Dix received an order from Washington to arrest their editors and proprietors and suppress the newspapers. Although the editors were shortly released, the order to suppress the dailies was not read to them until 22 May, four days after the details were fully ascertained, and only after the confessed culprit was behind bars.[103]

Should this course taken by the administration have appeared inexplicable, even stranger actions were taking place in Washington and points north and west. Early in the afternoon of 18 May, federal forces seized the Washington offices of the United States Telegraph Company. The United States company, commonly known as the "Independent," was a new concern established during the war in an attempt to compete with the government-supervised American Telegraph Company.[104] Thereafter, orders were issued to seize the Independent's facilities in New York, Baltimore, Philadelphia, Pittsburgh, and Harrisburg. The employees were sent to prison, and the office files were ordered searched and held.[105] On the afternoon of the same day, Henry Villard, leading spirit of the upstart news agency that competed with the AP, was arrested in Washington. His partner, Adam S. Hill, was kept "under observation," and Horace White, an old favorite of Stanton, was questioned in a long session by the secretary of war.[106]

The key question concerning this entire sequence of events was asked by the *New York Daily News:* "For what crime have these offices been thronged with soldiers?" The New York editors, after all, had published a fabrication, not a state secret, or so at least it appeared. Americans pondering with the *Daily News* the rationale for the action against the New York papers, Villard's agency, and the Independent Telegraph Company were unaware of two important sets of circumstances that conditioned Washington's response to the bogus proclamation.

Of these, one was a sheer coincidence that turned the entire incident into a farce. For on the night of 17 July two Americans stayed up, pored over their desks, put pen to paper, and composed a proclamation calling up hundreds of thousands of men for active military service. One was Joseph Howard, in Brooklyn, New York; the other was Abraham Lincoln, in Washington, D.C.[107] Viewed from New York the proclamation published the next morning in the *World* and *Journal of Commerce* was a local fabrication. Viewed from Washington, D.C., however, it was a state secret, leaked in Washington and transmitted by telegraph to make it to the first morning edition in New York, bypassing censorship in the process. This scenario, construed in Washington, implied serious crimes committed by a wire service that was indiscreet enough to expose the sensitive secret, a telegraph company that evaded censorship and transmitted contraband information, and newspapers that gave aid and comfort to the enemy by publishing it. For these crimes they were "thronged with soldiers."[108]

It took a day of confusion and some hard evidence before Washington was prepared to drop its plausible, but false, interpretation of the source of the bogus proclamation. Remarkably, the administration persisted in its original course of action. The New York newspaper establishments were not restored to their proprietors before four full days of public discourse on disloyalty, the proper limits of freedom of the press, and harsh administrative action. Nor was Henry Villard released before 21 May. But observers noted that after the belated release, "some choice scraps of news . . . found their way to the office of [his news] syndicate," courtesy of the War Department.[109]

Sorting out the affairs of the Independent telegraph lines pursuant to the bogus proclamation incident took even longer. Its properties were not released until the night of 23 May.[110] This was after the president of the company, John J. Speed, summoned to Washington, had convened with Stanton. When the lines were finally restored, "with a clean bill of health," they were independent in name only. Stanton had the company connect its wires to the War Department, subject to the iron grip of Colonel Edward Sanford, "military supervisor of telegraphic messages throughout the United States." The Independent's message traffic thus became subject to rigorous censorship and close control, but it was also expanded by a share of War Department business.[111]

The AP was not touched during the course of these unusual proceedings. Although suspicions of wrongdoing could have been attached to it with greater plausibility than to any other party, it apparently was never even questioned.[112] Furthermore, on several other occasions during the war the AP was responsible for disseminating throughout the country information that was positively injurious, yet nothing like the measures in the case of the bogus proclamation was ever taken against it.[113]

The bogus proclamation episode apparently provided the secretary of war with some fresh insights into the complexities of news control, and also with an opportunity to apply this education to the wartime information environment. For if in the short run the arrest of innocent men, the closing down of blameless establishments, and the flagrant violation of freedom of the press were an embarrassment to the administration, the situation as a whole was not bereft of some significant potential advantages. Edwin Stanton could have been counted on to realize this potential, to the fullest.

On the eve of a fateful presidential election, the bogus proclamations gave the secretary of war a golden opportunity to move against independent actors in the wartime information business. It was an excellent pretext under which to subdue the anti-administration New York newspapers, the Independent Telegraph Company, and the opposition news agency. Indeed, through Stanton's drastic moves Villard was tamed, and the opposition telegraphs were brought within the sphere of the administration's information control.

The strong feelings raised by government harassment of the *Journal of Commerce* and the *New York World* reverberated throughout the political campaigns of 1864.[114] The public was conditioned by pre-telegraphic traditions to view its First Amendment rights as being vested in newspapers. But it had no such sensitivity to government interference with more modern and less tangible media, such as telegraphs and news agencies; they could be controlled with impunity.[115] Throughout the war there was hardly a murmur against the monopolistic control of wire news by the government and the major wire service. News from Washington, from Bull Run all along the road to Appomattox was centrally controlled by the government and a loyal Associated Press.

PART THREE

The News-Telegraph Monopoly at Work

7

"A Double-Headed Monopoly": The Associated Press and Western Union in Symbiosis

After the Civil War, enterprising editors of the various regions set out to reconstruct telegraphic news arrangements and adapt them to conditions of peace.[1] They did so quite belligerently. Regional conflict in the news business would escalate to deadlock, broken only by an all-out struggle—the press association war of 1866–1867. American telegraphy had to make its own post-war adjustments, leading to a restructuring of the industry. In the aftermath of the press association war and telegraphic reorganization, a new deal was struck between the NYAP and the regional press associations. This settlement was based on drastic changes in the relationship between the news services and the telegraph.

Yet the settlement of 1867 would prove to be no more than a preliminary adjustment to post-war conditions. Significant changes in the ecology of news and wires throughout the Gilded Age would force repeated modifications of the compact. These changes, some gradual and cumulative, others sharp and abrupt, were destined to undermine even the most fundamental aspects of press association structure, leading to a revolution in the last decade of the century.

Notwithstanding the concessions it had to make to individual enterprise among its members, the Associated Press appeared to emerge from the Civil War as strong and dominating as ever in the national news market. Yet war conditions produced signs of discontent among client newspapers, most markedly in the West and in the South. This discontent gave rise to the first attempt by clients to organize to improve their bargaining power with the national news monopoly.[2]

With obvious displeasure, but in obedience to the orders of his employers, Craig in October 1862 raised the price of the Associated Press news reports to the outside press by more than 25 percent. This followed shortly after a sharp increase in telegraph tolls levied on the midwestern press. "Some of the Editors supposing, very stupidly, that the Associated Press controlled the telegraph lines, wrote some very abusive articles against the Association," noted Craig in his annual report for 1862. Worse still, a general meeting of the western editors was convened in Indianapolis to evaluate the situation. Craig informed his principals that he had succeeded in placating the westerners, but he implored them not to carry the gouging any further.[3]

"Having had the most ample experience of war and peace between the leading papers of the country," stated Craig, "I should feel extreme reluctance to abandon the latter for the former . . . An opposition News Association, properly organized," he supposed, "would speedily and surely extend all over the country, to the very serious detriment of your Association." Craig reminded the members of the determined opposition they had faced between 1850 and 1855, when they first invaded the national news market, and concluded his report with the admonition: "What has been done can doubtless be done again."[4]

Indeed it was done, late in 1866. The insurgents were the same western editors who had convened in Indianapolis in 1862; they had since incorporated their organization as the Western Associated Press (WAP).[5] Rapidly developing in pre-war decades, the Western press, like the region it served, was greatly stimulated by the Civil War. The more mature western journalism became, the less adequate did the quality, the quantity, and the relevance of the Associated Press reports appear to the editors of the region. And the less satisfied they were, the more they resented the price of the Associated Press service and the arrogance of those who administered it. Believing that their professional performance in war years compared favorably with that of eastern journalists, the western editors were prepared to challenge the NYAP once the war ended.

The New York editors had to set their own house in order before they could address the trouble brewing in the West. The war brought the tension between competition and cooperation among them to a head. Those newspapers that showed an outburst of energy and enterprise in war reporting found it difficult to settle back into the conditions of prewar years. The war had broadened the gap between the rich and the less rich members of the association. The former saw the association's news as a starting point from which to develop their individual enterprises; the latter, who viewed the news gathered by the common facilities as the end product, expected the association to free them from the strains of competition.[6]

Dramatic changes in the telegraph industry brought the rifts between regions and newspapers into focus and ultimately produced the press association war of 1866–1867. The two great telegraph companies fared very differently during the Civil War. In the East the American Telegraph Company was oriented toward the north-south flow of telegraphic communications. Secession wrought havoc on the American Telegraph Company's system, pitting two antagonistic polities against each other at either end of its formerly well integrated network of lines. In contrast, Western Union flourished under war conditions. It was well positioned for the increased importance of cross-country communication. Whereas the Atlantic cable, the great hope of the American Telegraph Company, was put on hold during war years, Western Union's pet project, the overland telegraph to the West Coast, was completed in 1861. It strengthened the Chicago-centered, east-west-oriented system of Western Union considerably.

Once the war ended, the reintegration of the American's system was effected with surprising ease, yet the damage war had done to the American in its rivalry with Western Union was irreversible.[7] Western Union proved strong enough to absorb the United States Telegraph Company, a developing rival that established lines which competed with both major companies. Western Union was also in a better position to purchase and integrate into its system tens of thousands of miles of military telegraph lines priced for a close-out sale. By the spring of 1866 the balance of power in the telegraph industry leaned heavily in favor of Western Union.

Cyrus Field's failed attempt to lay an Atlantic cable in the summer of 1865 left no other course for the American but to accept the overtures of the Western Union. On 12 June 1866 a consolidation of the two companies was formally effected. A month and a half later, Cyrus Field's dream of an Atlantic cable was fulfilled at last, but the fruits of his victory were enjoyed by Western Union rather than the American Telegraph Company that he had established.

The supreme importance of foreign news to the American press made a major realignment in the news-gathering business inevitable once the cable was operating successfully. The *New York Herald* welcomed the cable with a barrage of news messages from Europe. Each word cabled to New York cost the newspaper $5.00 in gold (approximately $7.50 in greenbacks) for transmission, and the *Herald* paid at the rate of $40,000 annually just to maintain the European network of reporters it had established in anticipation of the cable.[8] The *Herald,* in obedience to Associated Press rules, offered its cablegrams to the other papers, at their share of the cost of transmission.[9]

The other New York dailies faced a dilemma: to refuse the cable dispatches would be to concede an undisputed leadership in news to the *Herald*, yet taking them would lead to financial disaster, as shown in Table 4. "*Herald* bills up to August 17th $3,000—send on the gold," telegraphed a worried editor on the *New York World* to its proprietor, Manton Marble, vacationing in Newport. In another of a series of desperate telegrams, *World* editors recommended to Marble: "Read your Bible, say your prayers, and be home for the next meeting of the Associated Press."[10]

When on 8 August the *Herald* sent the entire text of a long speech by the king of Prussia across the Atlantic at an astronomical price, the astounded New York dailies did not know whether to suspect the *Herald* of a hoax or to believe that James Gordon Bennett had struck gold.[11] Although the other papers did request the *Herald* to authenticate its cable bills before they paid for the dispatch, the *Herald*'s real intentions became clear.[12] It was using its enormous financial resources to force the other members of the Associated Press to relinquish their joint foreign news-gathering business. Supported by Raymond of the *Times* and probably also John Russell Young, now in charge of the *Tribune*'s news policy, the *Herald* formally asked to be released from the obligation to share cable news.[13] It intended to sell its cable news to the press of the entire country for a price each daily could afford. In this way, it would receive enough to cover, besides cable bills, the great cost of its news-gathering operations in Europe. Perhaps the service would also be profitable and might occasionally provide some choice bits of news for ex-

Table 4 Weekly Bills Paid by the *New York World* for Associated Press Report and Special Telegrams, before and after the Laying of the Atlantic Cable

	Associated Press		Specials	
Date	1865	1866	1865	1866
August	254	441	59	389
	110	381	53	207
	157	291	195	335
	170	485	111	353
September	195	836	118	308
	435	745	94	929
	84	1,373	22	4,250
Totals	1,405	4,552	652	6,771

Source: Undated memorandum in the Manton Marble papers, Library of Congress.

clusive publication by the *Herald.* The monopoly of foreign news was now severely threatened from within the NYAP.

An apparently aloof Craig let the association's members learn about and digest the new reality in American telegraphy the hard way; he had already made his own analysis of the situation and arrived at some far-reaching and firm conclusions.[14] Although the elder Bennett was a shrewd editor and probably had a better grasp of the news business than any of his partners, he still was primarily an editor. Craig, in contrast, was not a journalist but a news vendor. He immediately realized that the cable had abolished, overnight, the strongest foundation for the news monopoly he had established. Heretofore, the Associated Press members had control over the first release of European news to the public. With the cable in operation, businessmen no longer had to wait for the publication of a newspaper to learn the state of European markets; they themselves could use the cable directly.[15]

Whereas the *Herald* wanted to finance the cable news service by expanding the newspaper clientele that would pay for it, Craig believed that the mercantile community could more easily subsidize the cable news bills and provide greater profits to the news service. His idea was to establish a commercial news organization that would supply a continual flow of European price quotations and other important commercial news through a network of news rooms established in the major urban centers.[16]

But Craig considered his ambitious plan to be beyond the resources of an association of seven New York dailies. Furthermore, an effective commercial news service depended on a telegraphic circuit of national scope dedicated to transmitting the reports at frequent intervals even during the busiest of business hours. The role of the transmission agency in the scheme was thus too great to allow the news-gathering element of the system to dominate it. The natural patron of a commercial news system was the master of the wires, the Western Union Telegraph Company.[17]

But besides the problem of cable news, Craig was compelled to consider the altered position of the Associated Press in the domestic field. He understood that working in conjunction with a telegraph monopoly weakened the position of the wire service considerably. The Associated Press could no longer use the immense patronage it controlled to play one telegraph company against another; now there were simply no others. Nor could he threaten again to establish rival telegraph lines; there were no longer other lines to connect with, and creating a competing national telegraph network required more than mere press patronage.

Not surprisingly, Western Union saw things in very much the same light as Craig. The telegraph monopoly found the general idea of adding a news monopoly to its powers very appealing.[18] Besides, the company had a tremendous excess of telegraphic facilities, acquired through numerous amalgamations with companies that had competed with them over the same routes. Craig's commercial news bureau would put these superfluous lines to good use, and they could be available at any time for transmitting any quantity of telegraphic matter to the press of the country.[19]

By November 1866, Craig had completed his plans with Western Union. The telegraph company was to contribute its network of operators, located all over the country, to aid Craig in collecting the news. It was also to provide trunk lines on all the major routes for Craig's exclusive use. Although Western Union demanded a majority of the news company's stock, the administration promised Craig not to raise transmission prices above the going rates.[20] All Craig now had to do was to convince the New York editors of the inevitability of his plan. Whereas some editors lent a sympathetic ear, the majority were not quite ready to abandon without a fight the advantages of the national monopoly they held. Without the approval of the association, Craig proceeded, covertly, to make all necessary arrangements to inaugurate his service in collaboration with Western Union by the middle of November.

Somehow, Craig's scheme became known to all members of the Associated Press. Horror-stricken, the editors instantly dismissed him from their service with a barrage of denunciations. The premature disclosure greatly complicated Craig's plan of action, yet he proceeded to perfect his arrangements and promised to inaugurate his service by 22 November.[21]

In the meantime, the western editors were well informed of the trouble that was brewing in New York. A militant gathering of the western association's board of directors decided that this was their chance to liberate themselves from the yoke of the NYAP. Two aggressive western editors, the *Chicago Tribune*'s Horace White and the *Cincinnati Commercial*'s Murat Halstead, headed east with power to negotiate a settlement with the "New York press gang."[22]

Halstead and White did not appear overly concerned when their demands on the New York association were turned down flat. While in New York they had several reassuring meetings with Craig, who promised them that his service would begin shortly.[23] Nor were they overly intimidated by the New Yorkers' insinuations that their preferential contract with the American Telegraph Company, which Western Union was legally bound to keep, made competition impossible. They knew on good authority that Western Union's executive committee had formally decided to do "like business for like terms for all patrons of the same class."[24]

In a stormy meeting of the WAP in the Chicago Opera House on 12 and 13 December, a majority of the members, including all the leading newspapers of the region, confirmed the action of their executive committee.[25] The war between the East and the West had actually begun officially two weeks earlier, when the western press began to receive news reports from Craig's United States and European News Agency. This was a violation of the Associated Press regulation that its customers take no alternative telegraphic reports. By then the *New York World* had joined the western insurgents. Late in November the paper began to receive Craig's dispatches, and as a result the remaining members expelled it from the NYAP.[26]

Regional associations and individual newspapers aligned themselves with one or the other of the two contestants, and Western Union looked on, at first with satisfaction, as both sides emptied their coffers into the company's treasury. Overall, Craig and the westerners appeared to be doing well. Craig had succeeded in retaining the loyalties of most of his former employees and brought over to the new organization virtually his entire network of experienced news agents. In the most impressive coup of the contest, Craig succeeded in effecting an agreement with Reuter's agency for an exclusive exchange of news. He thus secured for his organization an invaluable advantage in the foreign news field.[27]

Yet a determined NYAP fought back valiantly under the leadership of James Simonton, its new general agent. Although the association succeeded in holding many of its customers, after almost six weeks of rivalry its resources and its prospects of winning the fight were diminishing. The New Yorkers were now ready to make concessions, and Craig no longer could hope to end the struggle with his agency in full domination of the field. On 10 January 1867, the WAP together with the *New York World* signed a cease-fire with the New Yorkers. Craig agreed to withdraw from the field of general news and concentrate his efforts in the commercial news business, under the aegis of Western Union.[28]

During the press war, Western Union did not give Craig the support it had promised, and it showed no interest in controlling the general news. A month after the contest Western Union decided to terminate all its connections with Craig and to receive from the NYAP information for the commercial news department he had established on their behalf.[29] There was no longer a place for a "Jacksonian" entrepreneur in a business world of large and powerful combinations. In short, Western Union decided to abandon Craig and leave general news in the control of the press associations. One reason for this policy change was a change in the management of Western Union. Jeptha H. Wade, the company's president, who had made all the preliminary arrangements with Craig, was stepping down from active management owing to failing health. His successor, William Orton, either was not fully informed of the situation or was opposed to the policy. There was good reason for opposition, too.[30]

Editors in New York, Cincinnati, and Chicago were not the only ones trying to adjust to the new conditions in the telegraph industry. Politicians and statesmen as well, who were rendezvousing in Washington, were grappling for the first time with the implications of a private monopoly operating on a national scope in the sensitive field of information. In 1866, Congress became involved in the first of many attempts to regulate the telegraph monopoly.[31] With the avowed purpose of combating Western Union, both Houses approved and President Johnson signed, on 24 July 1866, a law allowing any state-chartered company to construct telegraph lines along post roads and through public lands.[32] Later in the year, legislators looked to more direct measures to combat Western Union's monopoly. By the time the press association war was under way, Congress was considering a reform bill introduced by Illinois's Elihu B. Washburn that called for the government to construct telegraph lines in direct competition with Western Union.[33] Western Union apparently decided that the control of press news was a luxury it could not afford, given its delicate position with the national administration and legislature. Control of the country's news would appear to add considerably to its powers, which were already viewed as excessive by reformers in Washington.

As early as 1860, a telegraph entrepreneur noted that a peculiar characteristic of Western Union was that its leaders "profess an excessive deference to the press as an exponent of public opinion."[34] Western Union could not afford to antagonize the press by wresting control of national news gathering at precisely the time that measures to break up its telegraph monopoly were being considered in Congress.

The end of the press association war of 1867 marked the beginning of a new era in the control of American news. Ostensibly, the NYAP retained its position as a national news monopoly, yet its power had been eroded significantly. Its agreement with the WAP made the latter responsible for news gathering and distribution in its region. Following the conclusion of the press war, the New Yorkers were forced into similar agreements with other regional associations, such as the New England Associated Press and the New York State Associated Press. The New Yorkers thus came to depend for their domestic news on autonomous regional associations. More important, it was now the regional associations that dealt with the individual members. It was up to them to decide who would receive the news when, where, and for how much. Moreover, each of the regional associations contracted independently with the telegraph company for the transmission of its news.[35]

The position of the combined press associations was further dependent on the goodwill of a national telegraph monopoly. An uneasy alliance between the news monopoly and the telegraph monopoly persisted for another fifteen years. Western Union stayed out of the news business in return for the association's commitment to use WU facilities exclusively for all of its business as well as for that of its individual members. The Associated Press could no longer play one telegraph company against another to secure advantages. Another clause in the contract between Western Union and the western press association stipulated that the latter would "not in any way encourage or support any opposition or competing telegraph company." It was well understood by all concerned that "opposition" was to be broadly construed. In Western Union's early battles against government regulation and competition, the press stood firmly in defense of the telegraph company. It was a combined news-telegraph monopoly that now supplied America with its telegraphic news.

Forged in an all-out showdown, the settlement of 1867 reflected the real balance of power between regions and associations, as well as between the telegraph industry and the institution of wire-news service. For fifteen years after the press association war, the general features of the accord of 1867 served to structure the overall relationship between the press and the telegraph. After a serious confrontation in 1881–1882 and its subsequent settlement, a new equilibrium would persist for another decade. The relationship between the press and the telegraph during this time was not, however, a stable one. Its survival was due, above all, to the continued dominance of Western Union in the American telegraph industry. The company managed to overcome a series of ambitious attempts to chip away at its hegemony over the industry. This sequence of telegraph struggles had, however, significant effects on the course and nature of Western Union's press policy.

The forces that were to ultimately shatter the equilibrium in the final decade of the century evolved gradually in the years following 1867. The most significant of these forces was the changing structure of the national marketplace for news. For during the Gilded Age, patterns of news flow adjusted to the dynamics of inter- and intra-regional shifts in population, commerce, and political power. The rise of the West and, in particular, of the far West, was accompanied by the process of urbanization. Reduced to the general contours of the press business, the process implied a shift in the relative importance of eastern and western news and a significant expansion of the metropolitan press. These developments, in turn, put increasing strains on the regional balance between press associations as represented in the accord of 1867, and on the internal structure of each of the regional associations.

Corresponding to these changes was a process of adjustment to the post-war reality of bigger business. Public perspectives on the nature and effects of private-sector enterprises operating on a national scale were maturing, just as business leaders were learning to manage and control these new giants. American telegraphy and the nation's wire services helped educate legislators, administrators, and the public to the implications of the new order of magnitude of business operations. The movement toward government regulation of big business, which played an important role in concluding the press association war of 1866–1867, would increasingly affect the development of the institution of wire news. From within the orbit of growing business organizations, innovations in structures, strategies, and management practices reflected adjustments to the maturing world of large-scale operations.

Other, more specific changes in the immediate environment in which the wire services operated effected a constant fluidity in the matrix of press and telegraph arrangements. One such factor was the development of institutional arrangements for direct supply of commercial quotations to businessmen, bypassing the newspaper as a medium. Another factor was the massive extension of the submarine cable business, which brought about a significant expansion and cheapening of intercontinental communications and international news flow. Finally, the notorious volatility of the New York City press scene made for continued reshuffling of personalities, interests, and policies in the NYAP. The discord and clashes among the members of the dominant press association were to have important consequences. In fact, the first signs of instability were injected into the new wire-news arrangements by the recalcitrance of a dominant member of the veteran press association.

James Gordon Bennett had not entirely given up his determination to keep the *Herald*'s specials and, in particular, its cable news, to itself. Rather than "sending around" the specials to his colleagues in the NYAP for a share of transmission costs, he still hoped to peddle them for profit. Bennett, in a surprise move, now aligned himself with Daniel H. Craig. Craig, bitter over being "thrown overboard" by the "telegraph folks" of Western Union, was again organizing a news company. In the background, leaders of the Western Associated Press were closely observing the New York situation. They realized that a potential competitor would keep the NYAP honest in carrying out the obligations it assumed in the settlement with the western press.[36]

This time it was the *World,* reinstated in good standing with the NYAP, that led the fight against the opposition. After Bennett, in an abortive attempt to force the hand of his partners, gave official notice on

29 June 1867 of the withdrawal of the *Herald* from the New York association, the *World* insisted it should not be allowed back.[37] The paper found a formidable partner for its campaign against the *Herald* in the Western Union Telegraph Company. Properly exercised over Bennett's move, William Orton casually reminded the NYAP that its contract would expire on 1 October, and that thereafter the telegraph company could dictate new terms. He also found it necessary on 9 October to notify the association that Western Union intended to withdraw from the joint venture in commercial news vending. This enterprise, originally created by Craig, added some $24,000 annually to the NYAP purse. These threats, coupled as they were with a secret reduction in the price of *Herald* specials issued by Orton, smoothed Bennett's retreat. Yet the retreat was merely tactical. The *Herald* went on growling at the telegraph monopoly from its editorial columns throughout 1867, and after regrouping it was ready to renew the fight.[38]

Early in 1868 William Orton noticed a sharp decrease in his company's receipts from its largest single New York customer. Yet the pages of the *Herald* were teeming with telegraphic news reports. They were clearly being transmitted over the wires of emerging competitors: the Franklin Telegraph Company serving New England and the northeastern corridor, and the Bankers and Brokers' developing system of lines, pushing west. On 28 February, Bennett withdrew all his special business from Western Union.[39]

More than the *Herald*'s news-gathering business was transferred to the opposition. In 1867 Bennett had established an evening newspaper, the *Evening Telegram.* The paper had the benefit of the Associated Press wire reports, yet Bennett thought that because the *Telegram* was nominally distinct from the morning *Herald,* it was not bound by Associated Press rules. The *Telegram* began selling Bennett's specials to a growing number of newspapers in New England, Philadelphia, and Washington. This emerging wire service was anchored on the Franklin Telegraph Company's expanding network of wires.[40]

To quell the new disturbance, Orton found it necessary to flash Western Union's ultimate weapon. He casually requested the Associated Press to grant the telegraph company permission to add a digest of general news to the commercial reports it provided to subscribers. The New York editors, Bennett included, well understood the implication—Western Union was proposing to enter the telegraphic news field on its own account. Grudgingly, Bennett apparently transferred the nucleus of his competing wire service to his friends at the Franklin telegraph.[41]

If Bennett's antics were not trouble enough for the telegraph company, it faced a significant challenge in the western theater of news operations. W. Scott Smith, Washington correspondent for the influential

New York Evening Post (an NYAP client), had been developing a telegraphic news service for papers outside the Associated Press, east and west. Smith supplied his western clients with news through the Bankers and Brokers company.[42] Late in 1867 that company was laying new wires in an ambitious extension plan with the crucial connection in Chicago as its main objective. By 1869 Smith gave his western business to what threatened to be a major western competitor, the Atlantic and Pacific system of railroad telegraphs. That company's network was based on the Union Pacific, just entering into a stage of rapid development.[43]

By early 1869, the entire field of news arrangements appeared to be in a state of turmoil. In New England, dissatisfaction with the NYAP service was widespread. Not only was the NYAP losing customers to Franklin's news service and to Smith's agency, but a coalition of New England Associated Press "way-papers," those published in the smaller towns and cities between New York and Boston, was organized under the leadership of Samuel Bowles. It was energetically pursuing a new and better news service.[44]

The Western Associated Press too was growing restive. It had received official notice from the NYAP of the termination of their contract of 1867. Weighing its options, the western association was again considering alliances with other regional associations. It began to negotiate an agreement with factions in the New England association and with the Southern Associated Press.[45] As if to fill the spatial gaps in this map of press service disturbances, signs of unrest appeared in the far West. John Hasson, the former NYAP agent, was supplying telegraphic news to Henry George's newly established *San Francisco Herald* independent of the California association's arrangements. Other, more powerful customers were beginning to show interest in the service that Hasson was operating out of Philadelphia.[46]

Resenting his confinement to the minor league of news rooms, with scores to settle and with weapons to wield, Daniel Craig sensed it was showtime. Experienced in adding drama to the telegraphic news arena, Craig prepared for what would be his last significant appearance on the wire-news stage. He renewed his alliance with Bennett, throwing into the deal certain telegraph patents he had acquired for rapid, or "automatic," telegraphic transmission, particularly adapted to the sending of long messages.[47] Bennett, obviously impressed by the potential of the new technology to revolutionize telegraphic news transmission, enlisted the *Herald* for Craig's campaign. At the same time, Bennett demonstrated great interest in trans-Atlantic telegraphy, revitalized by the French cable and other developments. He also set out to arrange for a comprehensive ship news operation in the port of New York in competition with the NYAP's long-standing harbor news arrangements.[48]

Should all these prospects deliver, Bennett and Craig could attempt to reopen the now closed issue of the proper alignment of press, telegraph, and wire service.

Out west and in fascination, Horace White and Joseph Medill followed the drama unfolding in New York. Apparently, they had thoroughly enjoyed their previous bout with the New York association and savored teamwork with Craig and Marble. From a distance they encouraged a potential alliance of Craig and Bennett, the WAP, and the *New York World*. Here was the nucleus for a national coalition of dissatisfied clients of the NYAP.[49]

Rather than firing in all directions, a composed William Orton evaluated this complex combination of disturbances to his company's press arrangements. After careful analysis he selected what he considered the key targets of his counter-maneuvers. Neither Craig nor Bennett was among them. Orton did not believe in the feasibility of automatic telegraphy or in Craig's legal rights to the patents he claimed.[50] Henry George and his Philadelphia connection were brushed aside as no more than an unpleasant diversion. Nor were the opposition wire services, giving aid and comfort to competing telegraph systems, considered more than a potential nuisance. In fact, Orton thought that the opposition press service, as long as it was limited to a small number of relatively weak papers, actually damaged the competing telegraphs.[51] Reading the map of news and telegraph competition, Orton identified two sources of real trouble—one in Washington, the other in New York.

True, the Franklin, Bankers and Brokers, and in particular, the Atlantic and Pacific, were thorns in Western Union's side. Given the heavy overhead weighing down the expense side of WU's balance sheet, the income that these competitors siphoned off drained WU's marginal profit. Competition thus prevented the company from maximizing profits from its comprehensive national system, precisely the effect most of the competitors desired. Telegraphy was widely recognized as a natural monopoly, and competitors ventured to force Western Union into buying them out at a heavy price. Such buy-outs, paid for in cash and fresh issues of Western Union shares, shrank the company's capital, watered its stock, and ultimately made managers' annual reports long on explanations and short on dividend announcements. Yet with all the inconvenience, competitors could hardly hope to eclipse Western Union or even seriously compromise its dominance of the industry.

In contrast, legislative action could do both; and by nationalizing the telegraph, the government could even destroy the veteran telegraph company. These possibilities appeared to Western Union officers distressingly real. Since 1866, reformers had been showing a peculiar, almost compulsive interest in the affairs of the company. For large companies that dominated their respective industries were a novelty, and a dangerous one at that. As one scholar put it, they "seemed to be mutations, the consequences of some sinister tampering with the natural order of things." They seemed, moreover, "not merely economic entities but powerful new political forces which must be opposed in the name of American democracy."[52] In Western Union's case, such dangers appeared especially great. The company's business was information, and through its operations it could directly affect the mind of America, shaping what people knew and what they thought.

Western Union's supposed power over the nation's commerce and politics emerged as a leading theme in the arguments for government regulation. Bills proposing that the government purchase the corporation, that the legislature regulate its prices and business practices, or that the government establish competition to Western Union (the "Postal Telegraph") were universally justified by the crucial role of telegraphy in the flow of commercial information and political news. The suppositions that "the newspapers are completely in the power of the telegraph companies," and that Western Union was "the master of the Press," were considered justifications for moving against WU.[53]

To prove the contrary, Western Union needed the backing of the nation's leading newspapers and press associations. In his efforts to secure that support, Orton faced what he considered a major cause of the press association unrest of 1869. This was a bold attempt by the NYAP to force the regional associations to concede to it the power to contract directly with Western Union. This demand by the New Yorkers was the basis for its terminating its contract with the WAP and the source of the turmoil in the wire-news affairs in New England and in the South.[54]

In his attempt to alleviate the tensions stirred by the NYAP, and to secure the support of the press for the telegraph company in its struggle against government regulation, Orton was guided by one axiom and two propositions. Past experience made it clear that the *Herald*'s opposition should be taken for granted.[55] But further, Orton considered it imperative to prevent conflict between the WAP and the NYAP. Overt conflict between the two major press associations served by the same telegraph company would inevitably lead to one or both of the combatants blaming its discontent on the telegraph. And it was certain that these complaints would be magnified by the company's enemies in Washington. Moreover, such a contest could result in one of the parties joining a telegraph competitor, to Western Union's detriment.[56]

According to Orton, it was also imperative that newspapers and the several press associations contract directly with Western Union for the transmission of their news. The telegraph company could then divide the press, the better to rule it; it could grant favors to certain papers and threaten others with sanctions wherever and whenever it became necessary. Moreover, were the NYAP to unite the press associations in their telegraphic contracting under its direction it could force Western Union into concessions by threatening to transfer business to the opposition. Worse still, it would have the ability to realize that threat.

By the summer of 1869 Western Union succeeded in restoring order to its press relations. Flanking the NYAP, it concluded a series of agreements with the New York State, the New England, and members of the Southern AP, cajoling them with generous concessions in schedules and rates.[57] Magnanimously, Western Union offered the Western Associated Press all the facilities it would require to commence a national news service of its own. But with no potential clients for a competing service left in the field, this was no more than an empty gesture. The New York association, thwarted in its attempt to bind its client associations closer to itself, and reminded again of Western Union's interest in establishing a news service of its own, was willing to renew its *entente cordiale* with the Western Associated Press.[58]

The Committee on Post Offices and Post Roads reported negatively on the bills for telegraph reform. It found no evidence that "the people wanted to make a change" in the national telegraph system. The proposed reform was forwarded in the interest of press freedom, but the committee found that "the press does not support it."[59] On 21 June Orton could notify his representative to the WAP, Superintendent of the Central Division Anson Stager, that "our press relations in all quarters appear now to be established on a firm and more satisfactory basis than ever before." "I trust it will prove permanent," concluded Orton, hopefully but erroneously.[60]

Ostensibly, the peaceful settlement of the conflagration of 1869 restored the status quo. To some, this was an achievement, to others, a conspicuous failure. Members of the WAP must have realized that under the circumstances the association had performed poorly. They had had a golden opportunity to improve their position but had missed it. Western Union, put on the defensive by Congress, was making concessions to the press, yet the WAP gained little. The NYAP was put on the defensive by WU, by its client associations, and by internal discord, but here too the West made no headway.

The western association was facing a problem of management. Its unique structure made rapid, decisive action in an emergency virtually impossible. A cooperative of more than one hundred newspapers spread over a very large territory, it had to delegate its powers freely; to the extent that these powers were conferred on the board of directors and the executive committee, they could hardly be exercised efficiently.

The elected officers were all active journalists busy managing their papers, and they lived a great distance from one another. Indeed, in the thick of the events of 1869, WAP action was no more than the improvisations of the *Chicago Tribune*'s Joseph Medill. Depending for information on his friends Manton Marble and Daniel Craig, he tried to close complicated deals in one hurried visit to New York.

In contrast, William Orton was able to play his hand well informed and well advised, receiving input via a continuous two-way flow of information and opinion up and down the ranks of middle management. Orton, as general manager, advanced the interests of his stockholders and the policy dictated by the directors and the executive committee through well-concerted action with his division superintendents and local managers, with the vice presidents (one in charge of press affairs), with the treasurer, secretary, and other staff officers, and even with the company's lobby in Washington. The NYAP orchestrated its moves more simply but at least as efficiently. The member-editors met frequently to decide policy and entrusted its execution to James Simonton, the general agent, whose actions were supervised by the executive committee. Simonton initially inspired little confidence either within the association or without, but shortly it became apparent that "for all practical purposes, he is the Associated Press."[61]

Leaders of the WAP realized that it was now time to adopt the innovations their friend Craig had bequeathed to the management of American business in general and to the wire services in particular.[62] They needed capable management endowed with power to run their widespread and complex operations. On 1 November 1869 the westerners finally adopted the logic of managerial capitalism and appointed William Henry Smith as their general agent. The résumé of this Ohian included editorial work on the *Cincinnati Commercial* and the *Gazette,* editorship of the *Cincinnati Chronicle,* service as secretary to Ohio's war governor, John Brough, and, later, to the state of Ohio.[63]

For a sometimes historian, devout religionist, and amateur politician, Smith made a surprisingly competent manager. And although he was something of a hypochondriac, he had a tremendous capacity for work. Smith appeared "small, wizard-like of countenance and with a hungry, cadaverous look" to an unsympathetic eye; he was ruthless in adversity, vain in victory, and self-pitying in defeat.[64] Of Puritan stock, he had a distinct self-righteous streak; he also understood power and craved it. Smith ultimately proved highly successful in his quest for power on behalf of his WAP superiors, perhaps even too successful. He would aggrandize the organization he managed to the point of counter-productivity to its principals.

At the helm of WAP operations, William Henry Smith proved extremely competent in attending to the details of the press service and in effecting efficiency, economy, and system in workaday operations.[65] By the summer of 1870 members of the executive committee could acknowledge with relief that they were no longer "troubled with the adjustment of multitudes of minor matters" or burdened by writing "reams of letters and . . . hundreds of telegrams." By 1871 they admitted that their labors had been "greatly lightened" by the "indefatigable efforts of the General Agent to keep all parts of the intricated and delicate machinery" of the association in good working order.[66]

The assistants Smith instated in his headquarters were soon handily managing the office routine, instructing the wide net of news agents, managing finances, corresponding with clients, and trouble shooting when necessary. With the day-to-day management running smoothly, Smith could devote his attention and energies to devising and carrying out strategy and policy on behalf of the association. His main objective was to secure as comprehensive a news service as possible, and to have that service attuned to the particular interests and needs of the newspapers the association served.[67] As Smith subsequently explained, his "rule has been *to give all the news,*" and this goal implied operating a "perfectly organized" news bureau. "This is thorough work," Smith offered in 1872, "and it requires a more thorough management than has yet been anywhere."[68] He would later be called "vast facilities Smith" in recognition of this incessant ambition for greater news facilities, preferably administered at a low cost.[69] The quest for vast facilities brought him into an early and bitter contest with WU and, after winning it, into close and harmonious working relations with its management.

Smith's campaign to adapt WAP news to western tastes and interests brought him into constant conflict with NYAP managers, whose news constituted an important part of Smith's WAP report. During the 1870's Smith stubbornly worked to expand the control of his New York agent, Charles A. Boynton, over the foreign news, Washington reports, and eastern seaboard items that the NYAP collected for its clients, to Simonton's great displeasure.[70]

Smith also worked to establish an infrastructure for a foreign news service that, in selecting the items for transmission, would be responsive to the particular interests of westerners. Due to the cost of cable transmission, the foreign news was the most expensive to gather, and its supply was an important source for the dominance of the NYAP in the coalition of press associations. Even with the gradual reduction of cable tolls resulting from the profusion of trans-Atlantic cables, parallel increases in the service made the bills for foreign news as high as ever. Throughout the 1870's the WAP paid the NYAP some $50,000 annually as its share of the expenses that the NYAP assumed in gathering the foreign news for the American press. Yet Smith believed that the association he managed could expand to a scale that would enable it to afford international news gathering independent of the NYAP.

Indeed, besides improving the association's news supply, Smith worked energetically to strengthen the WAP, increase its membership and clientele, and expand its sphere of influence and of territorial control. In this mission, he would encounter grave difficulties. There were definite limits to the potential expansion of the WAP's news business: geographical restrictions laid by the NYAP and restrictions on the penetration of its own territory laid by Smith's employers.

Smith coveted the rapidly developing regions lying on the margins of WAP territory. In some, such as Kansas and Missouri and the Northwest, the WAP had encouraged the formation of local press associations, tributary to the WAP. Smith resented this policy but was bound by it. It was formulated at a time when the association had considered efficient management and control over vast outlying territories beyond its capability.[71] Elsewhere, the WAP's *lebensraum* was constricted by the contract of 1867 between the New York and the Western Associated Press. When signing the contract, the westerners apparently did not anticipate the magnitude of the westward march of population and business. The agreement left Texas and the Southwest, Colorado and the Rockies, and the entire West Coast to the control of the New York Associated Press. Tensions between the two associations rose and fell during the 1870's in direct correlation to the intensity of Smith's drive to conquer the New Yorkers' western empire.

Smith was also severely handicapped in developing the power of the association within its legitimate territory. Shortly after the formation of the Western Associated Press in 1862, the NYAP conceded to its members the privilege of receiving the reports exclusively.[72] Apparently, this privilege was subsequently repealed by the New York Associated Press. One of the main goals of the westerners when they declared war on the NYAP in 1866 was to restore the privilege.[73] The victory of the Western Associated Press gave it the right to limit the sale of wire reports in its territory to members only.[74]

In the course of the late 1860's and 1870's, new newspapers were established to cater to rapidly expanding populations. When these papers were the first dailies in their communities, the association usually consented to supply them with a franchise in return for a cash "bonus." Where dailies with a wire service franchise were already established, it was up to the local members to decide whether to share the Associated Press report with the newcomer. When consent was given, it was usually in consideration of another large bonus to the existing newspapers.

In large cities, such consent was seldom given. This enabled Associated Press newspapers to increase their circulations considerably, without sharing the increasing business with newcomers.[75] By 1879 an Associated Press franchise in Chicago was valued at more than $100,000. In New York, the AP franchise was estimated at more than $500,000.[76]

Smith fought valiantly for the admission of new papers to the western association he managed, but he seldom carried his point with his principals. Unlike Smith, they were newspaper editors more than they were ambitious system builders. Ultimately, Smith would find a way to provide the outside press with telegraphic news, ostensibly without compromising the members' monopoly over AP reports. Smith entered on this course guided by more than personal ambition and professional pride, for powerful outside forces made it almost inevitable.

As the number of outside newspapers increased, a succession of competing wire services developed to supply them with news. First W. Scott Smith, as well as Bennett and the Franklin telegraph, then the Hasson agency, supplied the increasing demand. The Hasson agency eventually developed into the American Press Association, supplanted in 1877 by the National Associated Press; and a host of smaller organizations established networks of news gathering and distribution for the outside press. These organizations posed no serious threat to the Associated Press or to its member newspapers, because the service they supplied was much inferior to that of the veteran wire service. Moreover, all these attempts at competition were short-lived and ended up as expensive failures.[77]

Western Union was, however, seriously concerned with the emergence of opposition wire services. It was committed by its contracts with the Associated Press not to do business with competing wire services on more favorable terms. With fewer and less affluent newspapers to share the price of news reports, the opposition news agencies couldn't sustain their service having to pay Western Union the prices that the Associated Press could afford. They therefore gave their patronage to a series of telegraph companies that had emerged to compete with Western Union. The Franklin Telegraph Company, the Atlantic and Pacific, and later the Baltimore and Ohio and the American Union, together with other aspiring national telegraph systems, did a large business for the opposition wire services, to Western Union's chagrin.

By the late 1870's the problem of supplying a reasonable service to the rapidly increasing outside press assumed a new and threatening dimension. Supporters of government regulation of the telegraph had been active throughout the 1870's. The problem of Western Union was a mainstay on the national agenda, and since 1866 all Congresses but one had seriously considered bills to regulate the telegraph. WU's control over the national flow of news and over the American press were central aspects of the debate. But at first presidents, postmaster generals, congressional committees, and individual legislators, in an endless procession of addresses, annual reports, investigations, bills, and speeches, failed to determine the adverse effect of a telegraph monopoly on press freedom and on the public good.

But as information about monopoly of knowledge in America added up, the focus of the debate in Congress gradually shifted from the role of Western Union to the control the Associated Press exercised over the press of the country, over the flow of national news, and over the opinions of its people.[78] By the late 1870's the rationale for government action had been transformed. Regulation was no longer intended to free the press from the clutches of a telegraph monopoly; it was now aimed at protecting the people from a harmful press monopoly.

Discussing the necessity of telegraph regulation in 1879, the chairman of the committee of railroads stated that "what [had] done more to agitate this question and to call . . . for some legislation" was the "very great monopoly upon the part of the Associated Press."[79] By then, the government had learned all it ever wanted to know about the principles and practices of the Associated Press. It discovered that seven dailies in New York virtually controlled the national supply of news. It learned that this was done by having the nation's news "edited" or, to use a term that legislators were more comfortable with, "censored," in one office, by one man. It was told that Associated Press newspapers were not allowed to receive telegraphic reports from any competing news service. And it also found out that newspapers receiving the wire service's news were forbidden to criticize the Associated Press publicly, on pain of losing their franchise. The AP was the one institution in America that was immune from scrutiny by the press.

But the legislators' main discovery was that the privilege of receiving Associated Press reports was restricted; certain newspapers could get them, others could not. Supporters of government regulation of the telegraph industry at last found an instance in which the news-telegraph monopoly clearly resulted in a restriction of trade. Vested interests were preventing the establishment and hindering the success of new enterprises. Moreover, these restrictions applied to the publication of newspapers, one of the foundations of civil liberties. Legislators opposing the monopoly were cheered on by the outside press, its numbers and its influence rapidly growing.

With the Associated Press replacing Western Union in the hot seat of congressional investigations of news control, the balance of power between the two organizations was altered. Affiliation with the Associated Press was no longer an asset to the telegraph company in its efforts to combat government regulation; it was becoming, instead, a very dangerous liability. When in 1882 the New York Associated Press applied to the telegraph company for better terms, Western Union president Norvin Green, rather than conceding, lectured the president of the Associated Press. The association's refusal to supply the outside press with news, wrote Green to David M. Stone, "devolved upon this company, as its assumed ally, more odium and antagonism in legislative bodies, the popular mind, and the outside press, than any one thing in the history of this company."[80] WU was then searching for ways to relieve it of such odium and antagonism.

In the late 1870's all three major actors on the telegraphic news stage—the New York Associated Press, the Western Associated Press, and Western Union—were dissatisfied. By 1881 they became entangled in the same kind of triangular relationship that has been acted out a thousand times on the theatrical stage. The NYAP was jealous of, and felt threatened by, the growth and increasing assertiveness of its western client. As the western association expanded, it became more aware and more resentful of the constraints placed on it by the superior association. The NYAP realized that the WAP was cheating, supplying news out of its territory, in violation of the sanctity of their long-standing contract. Indeed, the WAP was selling news to the press of Colorado, Texas, and southern locations.

WU was party to this infidelity. Realizing that social and economic endowment tended to favor the West, it supported the WAP in a series of controversies with New York, and it covertly became party to ravaging the status quo by servicing the WAP in violation of its contractual obligations to the NYAP.[81] Before the telegraph company could pledge its fidelity to the WAP, however, it had to straighten out its own tangled loyalties.

In February 1881 WU bought out its two major competitors, the American Union and the Atlantic and Pacific, for $15 million and $8.4 million, respectively, in newly issued WU stock. Jay Gould and his group were large owners in both. The new stock, together with a significant block of shares acquired at low prices through a January bear raid on the exchange, gave Gould control of the telegraph giant. Exponents of telegraph regulation were elated. With Wall Street's Mephistopheles controlling the telegraph, the public, they believed, would finally realize that the "fountainhead of American news" was in real danger of contamination.[82]

Through the consolidation, Western Union fell heir to the business of the National Associated Press, serving the outside press at a very favorable schedule of rates stipulated by its contract with A&P. WU was now bound by that contract. Both the New York and the Western AP were seriously alarmed. The combination of WU's superior facilities and the low rates that the NAP was entitled to raised the specter of a competitor who could command vaster facilities than their own. James H. Goodsell, president of the NAP and the editor of the *New York Daily Graphic,* was an effective organizer and good newsman. He was perfectly capable of cashing in on the opportunity to outstrip the AP news service and then to lure the country's leading papers out of the AP. Responding to this contingency, the WAP demanded that WU honor its contractual obligation not to do like service at lower rates to other customers. The NYAP took the same opportunity to air the accumulation of complaints against the telegraph company for siding with other press associations, the WAP in particular. To make its complaints more persuasive, the NYAP notified WU officially in January 1882 that it was about to withdraw its business from the company and transfer it to the opposition remaining in the telegraph field—the Baltimore and Ohio, the Mutual Union, and the emerging Postal Telegraph Company.[83]

Western Union appeared to be in a quandary. Abrogating its contract with Goodsell would bring the NAP, the outside press, and inevitably, the nation's lawmakers down on the company. Honoring its obligations to NAP would alienate the country's major wire service and cause it to transfer its considerable business to telegraphic competitors. In response, early in 1882, WU commenced a series of secret moves. It began changing routes and rates on Goodsell, finding it impossible to provide the same circuits he had contracted for with A&P before the consolidation. Shortly, Goodsell was having difficulty servicing his customers. He was then approached by WU officers with confidential offers to drop his cause for a very generous bonus.[84] Yet Gould had no intention of allowing his telegraphic empire to become subject to shrill cries of monopoly and, inevitably, to government regulation. He would destroy Goodsell but also provide for the outside press, and that in a grand manner. Moreover, he intended to reap the profits for that service.

Gould's plan was no less than "to effect a revolution" in the press business. William Henry Smith was summoned to New York by his Western Union friends. Gould, together with top WU executives, gave him a grand welcome and went right to work soothing his ego.[85] They offered him an irresistible deal—forming a wire-news company led by the dream team of Gould, master of wires, and Smith, master newsbroker. Should Smith prefer public to private enterprise, Gould was willing to have the WAP stand with Smith under WU's canopy.

The proposed communion of press and telegraph had much to recommend it. Smith, with or without his western association, would give direction, press connections, and above all, legitimacy to the enterprise. Gould's dowry would be a national telegraph network for the transmission of news, its operators all to serve as news-gatherers for the new concern. His large interest in trans-Atlantic cables would make the costly challenge of international news gathering a manageable one.[86] Nor would the westerners' handicap in collecting New York and eastern seaboard news present a major problem—Gould would have his *New York World* collect that news on behalf of the new concern. Just as in 1866–1867, the *World* could serve as the westerners' Trojan horse. The combination would supply news to all the country's newspapers, to everyone's profit.[87] To lend legitimacy and civility to the plan, Gould together with E. D. Morgan offered the NYAP a third interest in a proposed tripartite combination with WU and the WAP. A majority of NYAP members found the proposition an insult, as Gould predicted.[88]

Although highly flattered, Smith was cautious. Moreover, he was loyal to his employers, who had given him full freedom and power to negotiate on their behalf. Smith decided to collaborate with Gould only as a representative of the WAP. Acting in the interest of the WAP, he demanded that Gould give up the clause in the proposed agreement providing for free trade in the new organization's news.[89] Also, Smith and his WAP backers wished to give an appearance of decency to their elopement with WU. The memorandum of agreement that Smith finally signed with WU on 22 June 1882 provided for three months in which the NYAP could reconsider its position before WU's and WAP's official notices to the NYAP of termination of contracts could take effect.[90]

It was now time for the NYAP, the isolated vertex of the triangular affair, to act. It had, however, become too ambivalent to move effectively. Formerly, James Simonton had succeeded in harmonizing the association's interests with the diverse concerns of its ownership. But in September 1881 he had stepped down from his position as general manager, owing to deteriorating health and increasing Pacific Coast obligations, leaving the association without leadership. Simonton's replacement, John C. Hueston, was a distinguished news-gatherer but a poor manager, blundering negotiator, and irresponsible policymaker. Given Hueston's weakness, the seven editor-members moved to the fore; and badly divided, they tried to steer the association's course.[91]

The NYAP newspapers were by no means comparable; their respective editors differed in their journalistic philosophies, in their politics, and at least as significant, in their attitude toward Jay Gould. The *World* was Gould's property and, perhaps, part of the *Tribune* too. The *Mail and Express*'s owner, Cyrus Field, was Gould's business partner. In contrast, the *Herald*, now under James Gordon Bennett, Jr., and the *Times* were his greatest detractors. They were the recognized ringleaders in one of the greatest, and perhaps most curious, crusades against an individual in the annals of American journalism. The *Herald* and the *Times* together with the *Journal of Commerce*—the three members most vehemently opposed to Gould, to the WAP, and to compromise—carried the day, with Hueston cheering them on.[92]

Busy inciting friends, antagonizing foes, and generally enmeshed in intrigue, Hueston did not notice the tightening of the siege around the NYAP as the summer of 1882 progressed into fall.[93] With WU squarely on its side, the WAP did not have to follow the usual course of negotiating alliances with the secondary press associations to bolster its opposition to New York. Quietly but effectively, William Henry Smith and his WU allies had completed all necessary arrangements to commence a comprehensive news service, national and international, independent of the NYAP. By October 1882 Smith was ready. He would put his national and international wire service into operation on 1 November.[94] This time, no territory would be safe from the western attack, not even New York City. The NYAP instantly grasped the seriousness of the situation, and all its members agreed to sue for peace.[95]

By then, Gould and WU had lost interest in their original intention of moving into the news business. Gould realized that no press association would join him in a plan that would diminish the protection and value of its members' franchises. And the public would not stand for WU's entering the business on its own. To keep competitors and congressional investigating committees from involving themselves in WU's affairs, Gould and WU attempted to achieve harmony between the leading associations. A restructuring of the relations between the WAP and the NYAP could effect peace, harmony, and revenue for WU.[96] The problem of the outside press was also solved without straining the relations of press and telegraph. Either with or without Gould's encouragement, the United Press was organized in June 1882, and it began providing news to the unenfranchised newspapers of the country.[97]

It was not long before the WAP and the NYAP, aided by WU's good counseling offices, renewed their liaison. The crux of the settlement, reached in December 1882, was equal power-sharing between the two major regional associations. The role of each association, within a unified news-gathering system, was once again defined geographically. But to ensure harmony between the associations, and to prevent disturbances from internal squabbles, the responsibility for the news service was vested exclusively in a joint executive committee representing both associations.[98]

The joint committee included five members, three representing the NYAP and two representing the WAP. The imbalance was due, above all, to conditions within the NYAP. Divided into three pro-Gould and three anti-Gould papers (the latter tagged "the ring"), the independent *Sun* would have to serve as umpire. The West was compensated for its under-representation in a body that was to reflect equal power by two provisions. One was that the *Tribune*'s Whitelaw Reid would represent the pro-Gould papers on the committee. Reid held stock in the *Cincinnati Gazette*, had a western background and sympathies, and had proved extremely friendly to WAP interests over the years. The other was the silent understanding that William Henry Smith would serve as the general manager of the consolidated news combination. With the establishment of the joint committee, Hueston was forced to resign; and to ease the transition, Erastus Brooks, the veteran editor of Field's *Mail and Express*, was appointed interim agent. In January 1883 Smith took command of the news business of the two associations.[99]

It took many years of constant acrimony and strife, and a great waste of resources, before the press associations adapted fully to the logical, monopolistic structure of the industry in which they were operating. Both telegraphy and wire news gathering and distribution, predicated on network, called for single, unified operations run by professional managers. Within months of his appointment, Smith effected vast improvements in the wire service, comparable only to the dramatic economies he had accomplished. Just as the bulk of wire news served to AP newspaper editorial departments doubled and even tripled, the business offices were receiving handsome rebates on the rates they had been paying for years.

Yet as long as there were outside newspapers to serve, and the UP to serve them with a duplicate system of telegraphic news, there was waste. To maximize efficiency, as well as economies of scale and scope, the two competing systems would have to be unified. Once the wire service was led by those whose primary responsibility was to administer the service and improve it, it was almost inevitable that cooperation between the two national systems would replace competition.

Indeed, in 1885 the AP's joint committee entered into a secret trust agreement with the UP. The agreement specified that the Associated Press would supply the United Press with news it acquired through its vast and expensive news-gathering network, while the United Press would share the great profits of its news service with the Associated Press.[100] From 1882 to 1891 the arrangement worked to everyone's satisfaction. Western Union transmitted the news for, and collected tolls from, two far-flung national wire services. The United Press profited greatly from selling a commodity that cost it little to procure. And the Associated Press enjoyed a large share of these profits. Each of the seven partners of the NYAP received substantial cash bonuses.[101] The WAP used its United Press bonuses mainly to lower the cost of the service for its members. The outside press was now well served, there was the appearance of competition in the news business, and reformers and legislators were relieved, assuming that the country was no longer threatened by an all-powerful, unopposed monopoly of knowledge.

A fundamental flaw, however, was built into the trust arrangement. Western Union and the United Press were model modern corporations, where directors and major stockholders dictated policy, which was efficiently carried out by a loyal middle management. In contrast, the Associated Press was a coalition of associations that shared power on a regional basis, and each had its own unique structure. The leading associations in the coalition, the NYAP and the WAP, were very different organizations. One was a partnership of seven New York dailies, the other was by then a cooperative of almost two hundred midwestern newspapers.[102]

When in 1891 the secret trust agreement with the United Press became known to all, the revelation did not greatly affect the NYAP. Its proprietors were actually joint stockholders who all knew about the United Press arrangement and benefited from it financially. The WAP, in contrast, was thrown into disarray. The members of the cooperative were both owners and customers. As customers, some were seriously damaged by the trust agreement, which allowed competitors to be supplied with the news they themselves had paid for. Some AP members had purchased UP franchises to prevent future competition, and then paid the UP for the news they originally financed as AP clients.[103]

The fundamental problem with the structure of the WAP was that its splintered ownership did not allow effective control of its management. Vested with unchecked power, management pursued a course of its own, which ultimately proved counter-productive for the owners. In 1892 the western members of the joint executive committee—Richard Smith and Walter N. Haldeman—were repudiated by their organization and had to step down. Smith was censured, although his services were still deemed indispensable. Whereas the NYAP wanted to continue its pooling arrangement with the UP, the WAP was against it, and a separation was inevitable.

Yet telegraphic news gathering and transmission continued to constitute what it was by its nature—a national monopoly. Either the WAP or the NYAP would have to give way. A protracted struggle between the two ended in 1897 with a complete victory for the westerners. After fifty years of leadership in American news gathering, the NYAP dissolved. The Western Associated Press, incorporated in Illinois as the Associated Press, secured a dominant hold on the flow of American news well into the twentieth century.

8

"Talking to All America": Power and Politics of the News-Telegraph Monopoly

Telegraphic news reports to the West in 1851 were meager, and Murat Halstead, a young daily laborer working for a Cincinnati newspaper, found a way to make the most out of them. The trains that arrived with the New York daily newspapers, when they happened to be on time, reached Cincinnati at 1:00 A.M. Halstead would wait for the train at the depot and snatch the package of exchange newspapers as soon as the mailbags were opened. Alerted by the telegraphic dispatches to the important stories, he usually had "the first slice" of news cut from the New York dailies ready for reprinting at 2:00. Halstead's efficiency, which at first he had to defend "against the accusation of extravagant exertion," was dependent on the telegraphic reports. "They had told me what to look for," he later acknowledged. Use of the dispatches to "stimulate the examination of the exchange newspapers [by] pointing out the quarters in which to look for details of interest" shortly became widespread.[1]

If in 1851 telegraphic reports from New York set the agenda for the nation's press, two decades later the AP was supplying the press with "everything, and something in addition."[2] In 1872, it was Murat Halstead, now a distinguished editor, who angrily accused William Henry Smith, general agent of the WAP, of "extravagant exertion" on behalf of the wire service. Smith's staff had just sent Halstead's *Cincinnati Commercial* and other western newspapers a seventeen-page account of the nomination of Grant and Henry Wilson by the Republican National Convention. Halstead had arranged for an exclusive report from the convention, but "the Associated Press rushed in and furnished quite as much as I wanted about the matter, if indeed not more than I would have cared to have obtained." The increased scope of wire service reports, argued Halstead, made a "uniformity in the papers," which he found "detestable." He pledged to reform the system, proposing to do it "even if it were necessary to break up the Association."[3]

Halstead's complaints were in vain, and the quantity of information sent by the wire service was further increased during the 1870's, making possible even greater uniformity in the press. Whereas each member and client of the WAP received about 750,000 words in 1866 and a little less than 3,000,000 words of report in 1868, in 1882 western newspapers received more than 7,000,000 words from the association, as shown in Table 5.[4] This was after WU agreed to lease wires to the AP for unlimited night service between the major news centers.[5] By the late 1880's the AP's news dispatch was comprehensive enough to constitute between 80 and 100 percent of news copy in midwestern small town dailies.[6]

Comparing the price of special news dispatches with the price of AP reports makes the increased reliance on the wire service understandable. In 1880, for instance, newspapers paid an average of only .075 cents for each word of the AP report, but 1.3 cents a word for specials; in other words, specials were on the average seventeen times more expensive per word than the uniform telegraphic reports supplied by the AP.[7] Not that WU could be suspected of philanthropy when it came to

Table 5 Scope of Western Associated Press Reports, 1868–69 to 1881–82 (Thousands of words/year)

Year	From East and cable	From the West	Market reports	Totals
1868–69	1,963	1,032	—	2,995
1870–71	2,159	884	390	3,433
1871–72	1,835	1,302	547	3,723
1872–73	1,834	1,370	590	3,793
1873–74	1,875	1,354	589	3,818
1874–75[a]	2,753	1,975	883	5,561
1875–76[b]	1,392	973	390	2,755
1876–77	2,072	1,291	561	3,925
1877–78	2,340	1,262	563	4,165
1878–79	1,958	1,285	570	3,813
1879–80	2,230	1,482	570	4,282
1880–81	2,568	1,487	470	4,526
1881–82[c]	3,128	2,011	2,100	7,239

Source: Western Associated Press, *Annual Report* (1869–1882).

a. Seventeen months (June 1874–October 1875).

b. Eight months (November 1875–June 1876).

c. Leased wire system introduced.

pricing AP reports. Although the wire service dispatches were relatively inexpensive for each daily, WU's income from their transmission was considerable. In the ten-year period between 1867 and 1876 WU's receipts for AP reports almost quadrupled, rising from $266,857 in 1867 to more than $1,000,000 in 1876. These dispatches, which constituted 4 percent of the telegraph company's income in 1867, were more than 10 percent of the company's receipts in 1876, as shown in Table 6.

The expansion of the service, and the increase in revenue to the telegraph company, were effected without a commensurate increase in the price of wire service dispatches to the newspapers.[8] Economies of scale and scope made this feat possible, owing to the vast network of wires WU commanded by 1867. With its wires stretching from coast to coast and reaching every populated region, WU could form very comprehensive circuits for transmitting wire service news. For instance, the entire press along the route from New York via Albany and Buffalo to Cleveland and thence to Chicago, St. Louis, Louisville, Cincinnati, and finally Pittsburgh received the AP report by one writing.[9] AP news was transmitted mostly at night, when there was little other business for the wires or the operators. Once the enormous circuits were connected, the reports could be made much longer with only a minimal increase in operating expenses. In contrast, special reports required the occupation of the wires by a single customer exclusively. Their price was therefore significantly higher.

The great increase in revenue for WU was made possible by the addition of new customers in the course of the 1870's and 1880's.[10] With the growth of villages into towns and towns into cities, new dailies were

Table 6 Western Union's Receipts for Press Reports and Total Receipts, 1867–1876 (in dollars)

Year	Press reports	Total receipts	Percentage
1867	266,857	6,569,000	4
1868	370,681	7,005,000	5.3
1869	404,906	7,317,000	5.5
1870	417,852	7,139,000	5.9
1871	493,085	7,637,000	6.5
1872	999,954	8,457,000	11.8
1873	1,041,080	9,333,000	11.2
1874	1,036,675	9,263,000	11.2
1875	1,051,898	9,565,000	11
1876	1,016,019	10,035,000	10.1

Source: Western Union Statistics, 1867–1876 (manuscript), New York Public Library.

established to cater to the rapidly increasing urban population. With the expansion of WU's network of wires, and with more places served by dailies being connected to the news circuits, the telegraph company's revenues for the transmission of a single wire service report could increase significantly.[11] Because the news could be dropped off along the circuits in even the smallest of the telegraph stations, at virtually no added cost, it was worth the telegraph company's while to furnish it even for a minimal consideration. Thus, a daily in Lima, Ohio, could receive exactly the same wire report as the great *Chicago Tribune* or the *Cincinnati Gazette;* but whereas the *Tribune* and the *Gazette* paid some $150 a week for the privilege, the telegraph company was satisfied collecting only $10 from the smaller paper.[12]

Not all cities and towns located along the route of WU wires could be included in the major national circuits. In relatively isolated population enclaves (such as in Texas and the Southwest or Michigan and the Northwest), where the march of the poles and wires reached a dead end, news reports had to be transmitted separately. In such cases, there was no justification for sending the full reports for what the newspapers—characteristically small and poor—could pay. The solution was to condense the wire service report into a much shorter dispatch (commonly referred to as the "pony" report), consisting of 1,500–2,500 words daily. The report was transmitted by special circuits to the outlying regions, and the newspapers there paid whatever the AP and WU believed they could afford.[13] Hence, even the most remote hamlet reached by WU's vast network simultaneously shared knowledge of the essential news with the great centers. With the domination of telegraphic news by two national giants, inequalities in the availability of news between regions, and within regions between urban centers and their hinterlands, were practically eliminated.

An important implication of the national circulation of uniform news was its tendency to promote and sustain a national market. Ever since the spread of the telegraph, commercial operators could learn of market conditions in a distant city in virtually real time. Indeed, the spread of the telegraph brought about a dramatic reduction in price disparities between regions and a sharp decline in intermarket arbitrage.[14] Commanding a national system of commercial news gathering, the AP was the one service that provided a consolidated digest of telegraphic data collected from all of the country's major markets. And because it served the newspaper press, the wire service greatly expanded the circle of those who transacted business guided by common knowledge of market conditions throughout the country.

By 1870 the NYAP supplied news to about 600 newspapers from coast to coast. Its report was to a large extent a digest of the previous day's commercial news. Thus, of the quota of 3,500 words it sent to the WAP daily (from 1867), it regularly devoted 1,600 words to eastern market news and quotations. The 2,500-word "pony" report sent to the southwestern press in the early 1870's usually included an 1,800-word summary of markets. Beginning in 1868, the WAP was allowed by its contract with the telegraph company a daily report of 8,700 words. About one half of this quota, between 4,000 and 4,500 words, was needed for commercial quotations.[15]

The western association, although it felt it needed much more general news, would not omit even one quotation from the list. Reader interest and newspaper demand for these quotations was much too great. In fact, the board of directors of the WAP informed the members in 1871 that there was "continuous pressure on the Executive Committee from members of the Association for more extended commercial reports, and they in turn are under the irrepressible pressure of their mercantile patrons."[16] Newspaper readers throughout the country could thus buy or sell and generally transact business with reasonably accurate knowledge of market conditions nationwide.

The AP also combined with WU to form a service that distributed commercial news throughout the day. For an annual consideration of $42,000, the AP supplied WU's Commercial News Department (the CND or "Commercial Bureau") with information that its agents procured by telegraph from all over the country, and by cable from all around the world.[17] In turn, the Commercial News Department supplied subscribers with a synopsis of prices in the major markets three or four times during business hours, and it also distributed special bulletins of important changes that had occurred, as reported by the network of AP agents. When it became necessary to transmit a commercial news bulletin, the WU network of wires gave it precedence, and for a few seconds was united into a national "CND circuit."[18] By 1873 the circuit included twenty-six cities, from Augusta to San Francisco and from New Orleans to Milwaukee.[19] The service made possible the equalization of information advantages for more specialized business transactions throughout the country and throughout the day. Indeed, with regard to markets, the AP helped realize Morse's vision of his telegraph's creating "one neighborhood of the whole country."

The wire service, by instant exposure of men and issues to national scrutiny, similarly transformed political communication; perhaps, as the editor of the *Cincinnati Gazette* Richard Smith claimed in 1881, it even "revolutionized the political world." The AP made it possible by then to print a 15,000-word political speech delivered in New York in the evening in the next morning's edition of the *Gazette.* And as a result, Smith pointed out, it was possible "to discuss the affairs of state at every fireside where the head of the family is not too stupid or too stingy to take a newspaper." What was emerging in consequence of expanding political reporting by the wire service was a national public opinion.[20]

Instead of "looking to Washington for political information," wrote Smith, "politicians look to the country for directions." Moreover, he argued, "The power that formerly lodged in a few men has been diffused," because

> when [a politician] speaks or acts he faces the whole people, and in twenty-four hours the great jury of American citizens pass upon his case. His utterances are flashed over the wires and printed within a few hours in thousands of newspapers and read simultaneously by millions of readers . . . Public men . . . are also careful of misrepresentation. There is no part of the country so remote that a public man can afford to make misstatements.[21]

Yet in the process of moving political debate from exclusive centers to the whole country, and from inside circles of politicians to "every fireside," the information first had to be filtered through the headquarters of the nation's wire services.

The function of the AP in the news industry may be likened to that of a giant funnel. From the vast network of wires of the nation's telegraph monopoly it gathered information from all over the country in its headquarters. There, the news was consolidated to a single report and, again through WU's vast facilities, distributed throughout the country. Not all information that reached those headquarters went out, and what did was necessarily transformed in the process.

Economic and commercial information, as well as other unambiguous news items, probably benefited from being centrally processed. In the case of political news, however, consolidation and standardization posed unique problems. Ideology, or at least political viewpoint, shaped the way events were perceived. The great political events of the day meant different things to different people, and it was, after all, people who managed the nation's major wire service and produced a single report for the entire country. They held the yardstick that measured news value and news definition of politically relevant information, and they sifted through it and edited it before supplying it to the country.

The news wire services, however, had always claimed to be impartial.[22] They professed non-partisanship in their attitudes, disinterest in the effect of the events they reported, and fairness in their mode of presentation. Maintaining neutrality was undoubtedly useful in the information brokerage business. It ensured access to information from all sources. Fairness in political reporting was also essential for obtaining and retaining the patronage of clients of all shades of opinion and varieties of political allegiance.[23] Professions of objectivity aside, however, it is still necessary to examine the actual performance of the wire service on this score. Bias in the selection of news items can be evaluated by studying wire reports within their political context.[24] Such a study of necessity requires an intensive investigation of a relatively short period. I chose to concentrate on political reporting by the AP in the late 1870's, for which period abundant sources are available.

The press of the 1870's was politically opinionated, even if less partisan than in former decades. Both politically affiliated newspapers and self-styled "independent" papers were expected to take sides in the great struggles between the two major national parties, and even to participate actively in intra-party skirmishes. The political activism of journalists and editors was most distinctly expressed in 1872 when they were the architects of the Liberal-Republican movement. They dominated its national convention and even nominated one of their own, Horace Greeley, as its presidential candidate.[25]

By the 1870's, management of the news wire services had shifted from professional news vendors to authentic newspapermen, who were characteristically political people. Moreover, the real power in the news organization was in the hands of a relatively small number of such newspaper politicos. The NYAP was closely supervised by the proprietors of the seven member newspapers. Its executive committee, therefore, exercised only limited power. Among the members, the *Times*, the *Tribune*, and the *Mail and Express* were staunchly Republican. The *Sun*, nominally Republican, was more flexible. The *Herald* was truly independent. Although on most issues the paper was in the Democratic camp, proprietor James Gordon Bennett, Jr., and managing editor Charles Nordhoff vacillated between the parties. The *Journal of Commerce*, generally less political than its six associates, had Democratic instincts. The *World* was the only one of the seven papers committed to the Democratic party. In 1876, however, it was sold to railroad tycoon Thomas A. Scott, and shortly thereafter passed into the hands of Jay Gould, whose political convictions were in accord with whoever was in power at the time.[26]

The general agent of the NYAP, powerful and independent, and in charge of supplying the whole nation with Washington news, was James Simonton, a Republican. He had replaced the apolitical Craig in the office of general agent late in 1866. Simonton had previously served as the Washington correspondent of the *New York Times*, and while its general agent in New York, he held a controlling interest in the *San Francisco Bulletin*, a Republican paper. Simonton was something of a stalwart, sympathetic to New York Republican leader Roscoe Conkling and his cohorts of party regulars; yet he seldom came into conflict with the more liberal proprietors of the association.[27]

In the WAP another small group of newspapermen exercised control. The executive committee chosen by the board of directors consisted of three men, the only members who oversaw the routine operation of the organization. As active journalists, engaged in managing their newspapers, the members of the executive committee did not supervise the work of the general agent very closely. The general agent's power within the organization therefore became paramount, and he could pursue a highly independent course.[28]

The two key functions in the organization were under Republican control between 1872 and 1892. Richard Smith, editor of the *Cincinnati Gazette* and a prominent Liberal-Republican, continually served on the executive committee. Joseph Medill and Horace White of the *Chicago Tribune* and William Penn Nixon of the Chicago *Inter-Ocean* alternated as representatives on the committee. The first two were influential in the liberal wing of the Republican Party. Nixon, while originally a "half-breed," supporting the faction of Maine's James G. Blaine and considered friendly to Republican reformers, became susceptible to stalwart influence after the financial salvage of his paper by friends of Senator John A. Logan in 1878.[29]

Walter N. Haldeman, proprietor of the Democratic *Louisville Courier Journal*, provided the executive committee with the appearance of regional and party balance from 1869 to 1890 but had very little real influence. From the Republican point of view, Haldeman was an ideal representative of the Democratic minority on the executive committee. His vociferous editor "Marse" Henry Watterson gave his paper the perfect image of an authentic representative of the Democratic South, yet Haldeman himself was far from a fire-eating southern Democrat. His affiliation with the Democratic Party was "an environmental rather than an ideological one," observed Watterson's biographer, and "his sympathies lay with the business interests of the country." Haldeman attended to his duties as executive with remarkable passivity, which enabled the general agent to pursue his course unchecked.[30]

The general agent of the WAP, William Henry Smith, wielded a great deal of independent power. Formerly he had been affiliated with Republican newspapers and later was elected secretary of state of Ohio on the Republican ticket in 1862. Smith was a regular party faithful and one of General Rutherford B. Hayes's closest friends.

The political preferences of the men at the top were reflected in the lower echelons of the organization. The head of the WAP's New York office, entrusted with supplying headquarters at Chicago with the bulk of eastern seaboard and Washington news (including NYAP reports), was Charles A. Boynton. A political ally of William Henry Smith, he was active in procuring private political information for Smith's personal intra-party campaigns.[31] Smith's underlings in the Chicago headquarters, William H. French and Walter Neef, who ran the day-to-day operations of the association, were his political confidants. All other WAP agents whose politics I have ascertained, including agents in Cincinnati, Louisville, St. Louis, Minneapolis, Milwaukee, and New England, were Republicans. In 1879, Smith positioned an agent in Washington in anticipation of the WAP arrangements to receive congressional reports directly from the Capitol. This agent's reports were so patently partisan that even executive committee member and staunch Republican Richard Smith advised the general agent to call him to order because such reporting would "raise a howl in the Democratic Camp."[32]

The predominance of Republicans in the management of the wire service made it easier for Republican politicians to gain access to the public. Henry V. Boynton, Washington correspondent of the *Cincinnati Gazette,* noted in 1891 the "singular fact" that "so few of the able men of the land" acquired sufficient knowledge of the press's machinery and of "the proper methods by which it can be used to promote their interests."[33] Boynton was referring to the potential nationwide exposure of items given to the AP. Until the practice of supplying handouts to the wire service became routine in the 1890's, the only way of securing the advantages of the national circulation of wire releases was through connections with AP personnel. Thus, all of some thirty requests for special coverage to be found in William Henry Smith's papers prior to 1881 were from Republican politicians with whom Smith was personally acquainted. These included Schuyler Colfax, William Dennison, Walter Q. Gresham, James G. Blaine, Carl Schurz, and James A. Garfield.[34]

The better access of AP newsmen to Republican sources did not result in exposure of Republican misdeeds. The AP records include much information about cases of Republican corruption that were never publicized by the wire service. The ethics of the profession at the time did not require journalists to publish all the information of public interest in their possession. In one case, the president of the WAP, two of the three members of its executive committee, and its general agent met to discuss evidence of corruption by a prominent Republican leader. Not one expressed the view that the news organization was obligated to expose the facts.[35]

The potential of the national wire service to influence national politics can best be demonstrated by its coverage of the nomination, election, and presidency of Rutherford B. Hayes. William Henry Smith, general agent of the WAP, was a close friend of Hayes's and became a member of his administration as collector of the port of Chicago. Smith was firmly committed to enhancing Hayes's fortunes. Thus, this case can hardly be considered representative. However, it illustrates the scope and limits of the wire service in manipulating public opinion.

In an early stage of the campaign to elect a Republican presidential candidate in 1876, the governor of Iowa wrote a friend that "Hayes of Ohio would do, but nobody out of Ohio knows anything about him." William Henry Smith undertook to correct the situation, and by the time the Republican National Convention met in Cincinnati, Hayes was a well-known name. In March 1876, John Sherman's warm endorsement of Hayes in a private letter to an obscure Ohian state senator was circulated nationwide over AP wires. Sherman's letter was followed in short order by expressions of support from two other Ohians of national stature—William T. Sherman and Philip Sheridan. In both cases the endorsements, made in private letters, received nationwide attention as a result of their circulation by the AP.[36] In contrast, Grant's expressions of sympathy for Hayes's candidacy (made after he realized that he had no chance of being nominated for a third term) were not relayed over the wires. Smith realized that they might alienate Liberal Republicans without enlisting much support from the stalwarts, who had in Roscoe Conkling a candidate of their own.

At the time Hayes entered the race, Conkling, Blaine, Oliver P. Morton, and Benjamin H. Bristow headed the long list of candidates for the Republican nomination. Blaine was the frontrunner. Hayes's nomination was possible only if the enthusiastic support for Blaine cooled before a majority of the delegates were committed to him. At the same time, the discord between the leading contenders would have to be great enough to prevent the formation of alliances among them. Once they were deadlocked, the nomination would go to an inoffensive dark horse who could unite the support of the major contenders. It was left to William Henry Smith, considered by Hayes's biographer to be the governor's "canniest advisor," to see to it that this unlikely set of circumstances materialized.[37]

Early in February, facts implicating Blaine in certain shadowy railroad security transactions came into the possession of General H. V. Boynton, a Bristow supporter and correspondent for Richard Smith's influential *Cincinnati Gazette.* Boynton believed that the Democrats were also in possession of these facts, and he assumed they were waiting for Blaine's nomination before making them public, so that the discrediting of Blaine would contribute to a Democratic victory in the national elections. It was obviously imperative for the Republicans to publish this information before Blaine could be nominated. Richard

Smith and Boynton presented Blaine with their findings but received no satisfactory explanation.[38] Willing to sacrifice their scoop to defend Bristow's candidacy, they forwarded the information to William Henry Smith for publication by the wire service, rather than publishing it exclusively in the *Gazette.*

Smith wisely held back. He informed both Joseph Medill, a WAP executive and a Blaine supporter, and Murat Halstead, president of the association and a Bristow supporter, of the accusations. Then he began employing his agents to gather evidence that would sustain the charges.[39] Medill, a warm supporter of Blaine, wrote him to ask for an explanation. In reply, the "plumed knight" assured Medill, "I should be a very great fool to permit myself to be placed on the parapet, to be raked by the Enemy's fire, if a single point of my harness was weak. I am Iron-clad as an oath." Yet he gave nothing by way of specifics.[40] In the meantime, W. H. Smith leaked some of the information to W. R. Holloway, editor of a Morton organ in Indianapolis and brother-in-law of the senator. Holloway was in possession of other incriminating evidence against Blaine, and he supplied it to Smith, hoping that the AP would publish the material.[41] Smith apparently went on to circulate Holloway's story confidentially among sympathetic Chicago editors, obviously hoping that one of them would publish it.[42]

Although the evidence against Blaine was accumulating, no editor risked publishing it, so the general agent called a meeting of those who knew the story. The two Smiths with Medill and Halstead, managers of the AP, met with Boynton in Cincinnati on 31 March. They decided to have Medill and Richard Smith send an ultimatum to Blaine demanding either a satisfactory explanation of the record or withdrawal of his candidacy. Otherwise, the WAP threatened to publish its suspicions before the Democrats had a chance to discredit a Republican presidential nominee.[43] Blaine, obviously stalling for time, evaded answering the charges; neither Richard Smith nor Medill received a response. The impasse was suddenly broken when Morton's supporters in Indiana, seeing Blaine's stock rise daily, lost their nerve and had some of the damning evidence published in a local Democratic organ. William Henry Smith immediately circulated the Indianapolis publication throughout the country, eliminating the frontrunner as the dominant contender.[44] Blaine's supporters placed the responsibility for the imbroglio on Bristow, whose supporters had uncovered the damning information, and on Morton, whose supporters caused it to be published. Neither Bristow nor Morton could now expect support as the second choice of Blaine's strong delegation.[45]

The decisive role William Henry Smith played in uncovering the facts, leaking them to the right quarters and leading the movement that made their publication inevitable, was never disclosed. Halstead, Medill, and Richard Smith, apparently trying to protect the general agent from gaining a reputation for involvement in partisan politics, never mentioned his role or even his participation in the Cincinnati meeting. As a result, Blaine remained cordial to Hayes.[46]

The discrediting of Bristow as a possible second choice candidate of the stalwarts was simple by comparison. William Henry Smith got word that Grant privately complained of having been treated discourteously by Bristow. The public, Smith believed, "do not want Grant for a third term" but "will not tolerate critical treatment of him."[47] Smith therefore privately circulated to his editor-friends Grant's recent complaint that Bristow and the treasury people "were aiming at me . . . They wanted to strike me." Grant's allegations, so damning to Bristow, received wide circulation.[48] Smith also sent an article to the *Chicago Tribune* exposing the dangers of a Bristow nomination, and he circulated it and the *Tribune*'s admittedly unconvincing editorial reply throughout the Northwest. "The two articles," Smith noted with satisfaction, "are provoking a good deal of active comment among newspaper readers."[49]

The AP rallied behind Hayes after he secured the Republican nomination and was pitted against Samuel J. Tilden, the Democratic governor of New York, in the national elections. Campaign documents supporting Hayes were widely circulated over WAP wires. The AP's most important service, however, was refuting anti-Hayes propaganda and preventing negative rumors from circulating. The most damaging of these alleged that during the Civil War General Hayes had been given money by one of his soldiers, to be remitted to the soldier's parents should the man fall in battle. The soldier was indeed killed, but the money Hayes received was never accounted for. The story was published in the ultra-Democratic *Chicago Times*, but before much damage could be done, Smith and the AP "flooded the Northwestern country . . . this morning and effectually killed the snake." The AP undertook a thorough investigation, which was widely publicized and totally exonerated Hayes.[50]

The WAP made it virtually impossible for Tilden's campaign material to reach the public through its service. The Democrats tried to compensate for the restrictions on their access to the public by circulating material through the "Literary Bureau," a New York-based organization managed by one of Tilden's secretaries that distributed feature material to the press. Later in the campaign, Tilden's pronouncements had to be sent out to the press as "specials."[51] Use of the AP machinery by Hayes's supporters was so effective that "emissaries of Tilden" apparently attempted to counteract it by bribing AP reporters. The newspaper record and Smith's testimony indicate that the attempted bribery failed.[52]

Smith expected that the October canvass in the midwestern states would serve as an indicator of the results of the national election and, moreover, that those results would influence voters. Anticipating a close race, he decided to cover the local canvasses personally for the AP, so as to make sure that "there is no mistake about the tone of the dispatches." But for all his fine-tuning of tone, the October results could hardly be construed to herald a Republican victory in November.[53] Indeed, notwithstanding the support of the AP, Hayes lost in the popular balloting by some 250,000 votes. Uncertainty over the final outcome prompted a fateful inquiry from Democratic national headquarters in New York to the *New York Times* editorial rooms as to the validity of the southern results. The *Times* office was frequented by its former correspondent James Simonton, the association's general agent.[54] The Democrats' questioning of the *New York Times* alerted Republicans, who hadn't realized how close the contest was, and considered their defeat conclusive, to the possibility of victory. This led to their claiming the election.

Both parties sent representatives to the South to try and control the actions of the local "returning boards." The close professional relationship between William Henry Smith and the officers of WU allowed the former access to Democratic dispatches to and from the South during the crucial stages of the count.[55] He could thus anticipate the movements of the Democrats and accordingly give instructions to Hayes's "visiting statesmen" in the South. Henry Watterson, one of the Democratic visitors, claimed in his autobiography that "the Republicans were promptly advised by the WU Telegraph Co. of all that had passed over its wires, [including] my dispatches to Mr. Tilden." William Henry Smith's correspondence fully confirms Watterson's accusation.[56]

During the debate in Congress over the proper legislative authority to supervise the count, the AP solicited and distributed legal opinions supportive of the Republican stand. At one stage, the Democrats chose the demonstration of public outrage at Republican frauds as their main strategy in preventing the seating of Hayes, organizing mass meetings in the Midwest to exert pressure on the legislature. The WAP simply did not report these meetings. To one agent Smith wrote: "I observe that the [Democrats] are going to meet and resolute on the sacred soil of Ohio . . . I suggest . . . that we can best subserve the interests of the country and of the press by making no reference to them in our reports."[57] At the same time the AP gave extensive play to statements of Democratic leaders who called for restraint and cautioned against mass disturbances aimed at preventing the seating of Hayes.[58]

Giving prominence to the opinions of one branch of the Democratic party in an effort to restrain its other factions was more than the Democratic press could stand, and they complained bitterly. In a characteristically self-righteous tone Smith lamented to the president-elect that "the Democratic newspapers have been abusing me in their customary brutal and unthankful way."[59] Yet the complaining did not last long. Writing to his friend Whitelaw Reid, Smith explained the Democratic attacks as "actuated by a common purpose to force me into consenting to the use of the wires in Tilden's interest." "They have failed," he assured Reid, and added, not so righteously, "If pushed to it, I could put the Democrats in a bad light before the country . . . They know it and are now getting in on my good side."[60]

The most important function of the AP in the post-election contest was promoting what became known as the "compromise of 1876." Serious historical controversy still rages over whether a bargain was struck between southern Democrats and northern Republicans in the aftermath of the election. Supposedly the former would help seat Hayes as president if the latter would end military rule and support internal improvements in the South.[61]

There clearly were, however, movements toward such an agreement, and the WAP figured prominently in them. It became involved in this Republican initiative because it was one of the few organizations operating on a national scale in which influential citizens of both the North and the South freely interacted. Another reason for WAP participation in the compromise negotiations was, paradoxically, the need for absolute secrecy. The active involvement of the most enterprising journalists of the day would leave none to search for information leaks. Moreover, having the North-South negotiations conducted by journalists as go-betweens, rather than by politicians of the opposing sectors, would not raise suspicion. As Smith summarily put it: "Such men as Richard Smith, Medill and [Andrew J.] Kellar could drop down [South] at once, quietly get to the bottom, and controlling news sources could keep the public in ignorance until the way should be cleared."[62] Not since a group of leading editors brought about Horace Greeley's nomination in 1872 had editors played such a prominent role in national politics, and it was the unique WAP structure that allowed for such effective involvement.

Arguably, the greatest challenge of the Hayes presidency was to restore or create an efficient and honest civil service to replace the chaos prevailing in government service after the Civil War, Reconstruction, and Grant's administration. The battlefield chosen was the notoriously corrupt New York Customs House, and the turning point of the two-year struggle between Hayes and Conkling's New York machine was the decision of Secretary of the Treasury John Sherman, early in 1879, to fully support the president.[63] Sherman's rather abrupt change of mind has heretofore seemed puzzling, but William Henry Smith's correspondence provides an explanation both necessary and sufficient. As collector of the port of Chicago, Smith was wrangling daily with Sherman over reform in the customs service. After losing faith in Sherman, Smith sent AP agents to find incriminating evidence of wrongdoing by the venerable secretary. Such evidence was indeed discovered, concerning intricate sugar import transactions. Once Smith came into possession of this evidence, a showdown was inevitable. "With a club for Sherman in one hand and [his] resignation in the other," Smith confronted Sherman, with President Hayes acting as arbiter.[64]

With the WAP at his disposal, Smith explained to a friend, he always had "this advantage that when the publication is made it shall be with our supervision."[65] Sherman, a sharp politician, recognized this great advantage. Realizing he was clearly outmatched, the secretary shortly produced his well-known letter declaring total war on the New York machine and support for the president.

Throughout the Hayes period, Smith gave the Democratic members of the WAP an effective lesson in information management. In turn, they "concocted a scheme to capture the Board for the benefit of the Democratic party in 1880."[66] The ever-informed William Henry Smith learned of this scheme and mustered enough Republican votes to prevent it. He thus was given the opportunity to control and manipulate information in the interest of his party for several more years. Rutherford B. Hayes once averred that "he who serves his country best serves his party best." His friend William Henry Smith proved the statement true, at least as it applied to serving the country with wire news.

Two queries and one observation remain to be made. If indeed the wire service was definitely biased in favor of one of the major parties during at least one period, how did the notion of the news wire service as objective become so pervasive? And then, how could an organization servicing the press of both parties afford to be so partial? The answers probably lie in the nature of partisanship, the nature of the wire service, and the nature of telegraphic news.

Some allowance must be made for the atmosphere and political climate of the 1870's. What appears partisan by our standards of objective reporting may have seemed relatively bias-free in an age when effective "waving of the bloody shirt" could win an election and any postmaster in the state of New York was ipso facto a Republican. Journalists were political people, and this was reflected in their papers. Newspapers reported the world from their corner in the political arena, and their partisan point of view was seldom concealed or understated. In comparison, the wire service did not explicitly take sides or express opinions in its reports; its bias was in its selection of news, not in its presentation. Against the backdrop of 1870 journalism, the news wire service could reasonably have been considered objective. Moreover, the kind of information the news wire service was intended to report was useful to newspapers of all shades of opinion. Because the service dealt in news events as they occurred—statements being made and letters sent—the customer newspapers could incorporate the news wire reports into their columns in any form they saw fit, or choose not to publish them at all. The public received the same telegraphic dispatch as part of a pro-Republican article in one paper and in support of a Democratic argument in another.

Editors viewing the AP reports from the perspective of newspaper-making were perhaps oblivious to the offenses of information management. The function of editors was to make a newspaper out of the best-written information that reached their desks. Smith and Simonton's task was to extract occurrences from the infinity of human events and send them to the editors. The editor, who did not see the whole picture, may not have noticed its coloring.

It was thus the monopolistic position of the AP in gathering telegraphic news that gave it the power to manipulate information. The AP was the sole operator in the news field that concentrated the entire national output of news. As a result, the AP alone could sin in omission and commission in sketching the diary of the American dream. Its monopoly of news distribution, in turn, secured its exclusivity as "the fountainhead of American news."

How representative was the Hayes period, with regard to information management by the wire service? In the *Brass Check* (1919), Upton Sinclair, in the best tradition of American muckraking, depicted a very Republican, big business and anti-labor AP controlling news in turn-of-the-century America. Closer to the Hayes period, the most serious accusations of AP political partiality came during the presidential election of 1884.[67]

By that time, William Henry Smith was joint general manager of the NYAP and the WAP. He operated from Chicago, and his political confidant William Henry French managed the New York office. In the absence of Smith, Whitelaw Reid, editor of the *Tribune,* the most partisan Republican newspaper in the city, looked after the operations of the AP machinery in New York. Reid, together with James Gordon Bennett, Jr., proprietor of the *Herald,* and Charles A. Dana of the *Sun* were then the executive committee of the New York Association. At the time of the election of 1884, Bennett, as happened so often, was somewhere at sea, and Dana was busy trying to compete with the skyrocketing circulation of the *New York World.* Joseph Pulitzer, the staunchly Democratic editor of the *St. Louis Post-Dispatch,* had purchased the faltering newspaper from Jay Gould the previous year and made it a real power in metropolitan journalism.[68]

Smith was not a great admirer of Blaine. Yet his old and dear friend Whitelaw Reid was, and Smith promised Reid to support the campaign "on account of my love to you." "I want you to win," he wrote Reid, "and to win a tremendous victory." Although that did not happen, they both apparently put forth their best efforts, using the considerable powers at their command.[69] Blaine, who was always cordial to Smith, wanted him personally to look after the telegraphic circulation of his campaign material to the press. Reid, and the more experienced Smith, made efforts to circulate campaign matter favorable to Blaine and to suppress unfavorable information.[70] But what they considered one of their greatest victories in these endeavors proved to be their greatest failure, perhaps even the worst blunder of Blaine's entire campaign. The AP made an unprecedented effort to give full and prompt coverage to a dinner in honor of Blaine, which would be his final major appearance in New York, the state in which—according to most observers—the election would be decided. The dinner of 29 October, organized by AP member Cyrus Field, featured the leading Republican capitalists of the city, who demonstrated, by the lavishness of the menu at Delmonico's and by the warmth of the toasts, their enthusiastic support for Blaine. Blaine's address at the dinner was a glorification of wealth. He pointed out to Jay Gould, Russell Sage, John Jacob Astor, and some two hundred other millionaires that it was the Republican party "which has enabled you to make this marvelous progress."[71]

The speechmaking ended that evening at eleven o'clock. Proudly, French informed Smith that the report of the dinner "was done so promptly and accurately that it was simply perfect." "Whitelaw Reid," wrote French, "was most gratified." The AP reporter was the only representative of the press allowed by the organizers to report the dinner. A complimentary copy of the proceedings was, therefore, given to ostensible competitors, the United Press and the New York City Press Association.

French sent the report from his office by 12:23, after allowing Blaine to make corrections of the draft. Blaine, however, failed to see that his whole speech had been a mistake. Eulogizing wealth when the great mass of voters were suffering from the severe depression was impolitic and unwise. French's admirable speed allowed the story to reach all the morning newspapers with plenty of time for editors and composers to give it all the attention it deserved.[72]

Indeed, Democratic editors gave what came to be known as "Belshazzar's feast" no less prominence than Republican editors. The next morning Walt McDougall's cartoon of "Belshazzar Blaine" sitting to feast with the "money kings," with a half-starved family looking on, hoping to salvage some leftovers, was spread across the front page of the *New York World.* It depicted what many other editors had to say of the affair, in many hundreds of words.

New York indeed turned out to be the pivotal state in the election, and Blaine lost it by 1,464 ballots, many those of family providers who had lost their jobs in the economic crisis. Blaine, Reid, and many other commentators ultimately attributed the defeat partly to the identification of Blaine with the money interests as represented at "Belshazzar's feast."

At first, however, neither the Republican candidate nor the AP conceded defeat. The count was very close, and the *Tribune* held out for four days before admitting that Cleveland had won. It was the AP's coverage of the count of New York's vote that gave rise to the gravest accusations of AP partisanship ever made. Here again, it was not the tone of the reports, but the manipulation and selection of information that was the issue.

The early tally of the votes was traditionally a joint venture of WU and the AP. The association's agents at regional centers compiled the results sent by WU operators from every hamlet and town with a telegraph office. The telegraph company supplied the headquarters of the AP with a continuous synopsis of the nation's polls.[73] Late on the night of the elections, the AP usually announced a preliminary result, and the following day results could be published with a high degree of certainty.

In the election of 1884 the New York vote and, therefore, the outcome of the national contest, were considered by the AP to favor Blaine. William Henry French even staked his reputation on a Republican victory, assuring Blaine in a telegram of 5 November that New York would go Republican by 10,000 votes. For three days thereafter, the AP tally either favored Blaine or was inconclusive. The count of every other source that collected returns, including the United Press, showed a victory for Cleveland.[74]

The *New York World* had what it considered a sufficient explanation for the hesitancy of the news-telegraph monopoly. The NYAP and WU had concocted a scheme to "steal the elections." By delaying the report of the final results and considering them too close to call, claimed the *World,* the AP alerted Republicans that the margin against the party was very slim. Given sufficient time, supporters in some predominantly Republican counties in up-state New York could steal enough votes to tilt the scales in Blaine's favor (results from rural western New York were the last to come in). If a Republican victory was initially claimed, results doctored by unscrupulous Republican managers would be received without doubt or suspicion. The *World* made similar accusations against WU regarding the announcement of returns in Indiana, another close state.[75]

The accusations appeared credible enough to draw thousands of New Yorkers to the streets. They paraded to WU headquarters and tried to break into the building, destroy the news-telegraph monopoly, and also, as their battle cry proclaimed, to "hang Jay Gould." Wisely, WU's president Norvin Green anticipated the action and alerted New York's police commissioner in due time. Through two days of demonstrations great numbers of the city's police prevented the mob from damaging the building, injuring its inhabitants, or even hanging Jay Gould.[76] After a conference with Norvin Green, Joseph Pulitzer and his *World* dropped the issue.[77]

Yet the case against the AP for trying to steal the election was not dropped by the silencing of Pulitzer. The conservative David M. Stone, for twenty-five years president of the association, announced to the public in his esteemed *Journal of Commerce* that "the election returns have been collected under the entire supervision of [Whitelaw Reid] the editor of the *Tribune.* The partisan character of the service is most painfully apparent and we do not believe that the other members of the association will quietly submit to lose the good name of the body in any such fashion."[78]

After the election, several editors in New York and New England served by the NYAP tried to establish themselves as an autonomous unit within the AP framework.[79] The insurgents were all Mugwumps, and the movement began with complaints over the AP performance in the controversial count of the election of 1884. During those "heated days of politics," Smith reminded his friend Reid, the Mugwump editors conspired to "stampede the entire press of the country and break down . . . the Joint executive committee [of the NYAP and the WAP]."[80] Those editors had come to realize the power of the wire service to influence national politics and wanted to use it for their own ends. Although the movement of the Mugwump editors ultimately failed, their "stampede" nevertheless had its effects. Their accusations that WU and the AP had tried to help the Republicans steal the elections of 1884 were not proved.[81] The public, however, was informed for the first time of such a possibility and came to realize the power behind the closed doors of the building on Broadway, on the corner of Dey Street, headquarters of the national news monopoly.

The news-telegraph monopoly, from its formation, operated in the shadows, hidden from public scrutiny. The elections of 1884 helped tear the veil from the AP and make it more accountable to the people who depended on it for their news.

"The press is not immaculate," acknowledged Henry V. Boynton, yet "the work of no other [profession] is so open to publicity. There is no veil, as there is for all others, under which the press can hide its shortcomings. The sunlight by day and the search-light by night illuminate the paths of all its known workers."[82]

Conclusion

> Along the smooth and slender wires
> The sleepless heralds run,
> Fast as the clear and living rays
> Go streaming from the sun:
> No peals or flashes heard or seen
> Their wondrous flight betray,
> And yet their words are quickly felt
> In cities far away.
>
> Nor summer's heat, nor winter's hail,
> Can check their rapid course . . .
> They rush, with news of weal or woe,
> To thousands far away.

The *New Orleans Daily Picayune* noticed a "magnetic neatness, so to speak," in the lines of this poem of 1847. From the perspective of a later period, however, some signs of an uneasy transition from perceptions and images of the age of transportation to the age of electric communication are apparent in Reverend J. Lyons's "The Magnetic Telegraph."[1]

Though justly impressed by the speed, continuous operation, and dependability of telegraphic communication, and acknowledging the revolutionary fact that information could be moved without visible signs of movement, the author could not fully comprehend the divorce of communication from transportation. It was still thought that information required "heralds" to convey it. The process by which one news message could reach thousands in many distant cities simultaneously appeared "wondrous," a performance equaled only by the rays of the sun. Only in later years were the implications of these revolutionary departures more fully apparent.

In 1844, point-to-point telegraphic communication became available to the American public with the construction of a government-sponsored telegraph line between Washington and Baltimore. Morse proudly reported to Congress, two weeks after the opening of the line, that it had already demonstrated its great usefulness. A Washington family heard a rumor that one of its members had been killed in Baltimore. By means of the telegraph, the family was relieved in a matter of ten minutes "from the state of distressing suspense." Similarly, a Baltimore businessman confirmed, by a short telegraphic inquiry, that a check handed to him by a Washington customer was good and represented funds in the Washington bank on which it was drawn. Morse illustrated by these examples "the immense advantages of such a speedy mode of conveying intelligence," in social life and in business.[2] The *New York Tribune*'s "Mercature" pointed out the potential usefulness of the invention to the political process by suggesting that if "heretofore, the District Representative had voted and acted conclusively on many important measures, without the advice of trusting constituencies" (whom Mercature assumed, perhaps naively, "he would have gladly consulted"), henceforth, through the use of the telegraph, "all important legislative measures will at once be placed before the people by a sleepless press."[3]

In the summer of 1846, the pioneer telegraph lines stretching north from Washington, south from Buffalo, and west from Boston connected in New York and were thus transformed into a network. All points along the previously isolated lines could now receive the same news simultaneously, by a single transmission. With the emergence of a telegraph system, Morse's early notions of the potential usefulness and impact of the telegraph had to be modified. Far beyond the ability to speed bilateral exchanges between individuals and institutions across space, the telegraph provided a whole new dimension of communication. This was the potential power to centralize, even nationalize, information. Employing the telegraph as a one-way means of transmission, a single sender could have his message reach all points on the telegraph network simultaneously, and thus could impress "the whole nation . . . with the same idea at the same time."[4]

The *New York Express* was quick to notice the significance of the added dimension of networking to the functions of the telegraph. It proposed, as early as 1846, that the "intensity and rapidity of communication" by telegraph would transform American society and politics. "The Union," suggested the *Express,* "will be solidified at the expense of the State sovereignties." It predicted:

> We shall become more and more one people, thinking more alike, acting more alike, and having more one impulse. Washington is as near to us now as our up-town wards. We can almost hear through the Telegraph, members of Congress as they speak. The country now will be excited from its different capitals. Man will immediately respond to man. An excitement will thus be general, and cease to be local. Whether good or ill is to come from all this, we cannot foresee, but that the Telegraph is to exert an important bearing over our political and social interests, no man can doubt.[5]

It took more than a decade for the fledgling telegraph industry to consolidate the separately owned lines into large regional companies controlling unified operating systems. The industry was prompted in this direction by its single largest customer, the New York Associated Press. A partnership of six New York dailies established in 1846 for procuring telegraphic news in common, the Associated Press represented the most significant effort to employ the telegraph as a national network. As early as 1851 the association, under the leadership of its general agent Daniel H. Craig, had established a system for centralizing the nation's news output in New York and supplying it simultaneously, in a consolidated report, to newspapers all along the country's telegraph routes. By 1856, the consolidation of numerous telegraph companies into a few major conglomerates made it possible for the Associated Press to operate a smooth working system, national in scope, for diffusing public information to the press.

"Processes of communication are the basis of the coherence of societies [and] cultures," suggests the scholar Karl Deutsch. "Even where we have one people," he adds, "the range and effectiveness of social communication within it may tell us how effectively it has become integrated, and how far it has advanced, in this respect, toward becoming a nation."[6] The expansion and consolidation of the telegraph industry, and the emergence of the Associated Press as a dominating national news supplier, indicate that America advanced considerably on the road to "becoming a nation" in the course of the 1840's and 1850's. Dramatic reductions in price differences between the country's leading markets, and the integration of regional markets, serve to confirm this notion.

It was commonly assumed in the nineteenth century, and later as well, that improved communication capabilities facilitated political unity. A conversation of the 1890's between Associated Press director and editor of the *Atlanta Constitution,* Clark Howell, and Charles Diehl, the assistant general manager of the Associated Press, expressed that assumption explicitly. On the porch of the state capitol building in

Montgomery, Alabama, the site of Jefferson Davis's inauguration as the president of the Confederacy, Howell tried to answer Diehl's query on how a civil war was at all possible. "The North and the South . . . did not know each other," Howell responded. "If the Associated Press had then existed," surmised the former Confederate, "and the news of all sections of the country could have been transmitted and read, it might have, and most likely would have prevented the conflict."[7] The only problem with this counter-factual assumption was that the Associated Press had, in fact, existed for fifteen years prior to the firing on Fort Sumter. During that period it supplied news of the North to numerous southern papers (which commonly accused it of serving as a "Black Republican" mouthpiece), and southern news to northern papers (which usually found a clear pro-southern bias in the reports).

As it turned out, in parallel with the development of a "tightening net of communications" in the 1850's, political unity in America was actually disintegrating in a process that ultimately escalated into civil war. Obviously, to the extent that telegraphic communication affected public opinion and public policy, it had hardly succeeded in fostering national unity and, perhaps, had even worked to upset it.

The American experience of a breakdown in social and political unity following the introduction of improved means of communication may reflect an example of what Lewis Mumford considered "the paradox of communication." Greater and more intensive interaction between individuals, sections, and nations, Mumford argued, can serve at first to "increase the areas of friction," to "hasten mass-reactions," and perhaps, more generally, to highlight disagreements and differences that previously were only latent.[8]

This potential for divisiveness as a result of rapid mass communication had also more specific and direct implications, especially with regard to political communication. In the summer of 1858, Confederate President-to-be Jefferson Davis may have gained some insight into the power of the new media to disrupt traditional political discourse. Ordered by his physicians to find refuge from the heat of his native Mississippi in a cooler climate, Davis and his family headed for Maine early in July. Moderate weather notwithstanding, political tempers in the North were just as heated as in the South. Yet Davis and his Yankee hosts found, to their mutual surprise and delight, much common ground. National news media had served to separate the political debate from the debaters. Jefferson Davis and his northern audiences found that once face-to-face, person-to-person communication was resumed, conflicting statements and opinions did not appear too threatening, or even irreconcilable. Physical proximity and the possibility for an exchange of ideas and sentiments put a different appearance on political disagreement.[9]

At precisely the same time that nonmediated personal debate was celebrating its power to foster harmony in the North, events in Mississippi demonstrated clearly the divisive potential of instantaneous communication. Throughout Davis's sojourn in Maine, and especially during his trip home through Massachusetts and New York, those "sleepless heralds," the dots and dashes of the telegraph, were conveying to Mississippi and the rest of the South a precise account of his utterances and the enthusiastic response they elicited in the cities along his route. Suspicions that Davis was compromising his positions and betraying the South became deeper with each new dispatch. The reports reached the South so far in advance of Davis himself that by the time he arrived in Mississippi public outrage had built to a very high pitch. Davis had to face an excited session of the legislature in Jackson and plead with his colleagues to take his word and his pledge of loyalty against the adverse reports in the press. Clearly, the fact that messages reached the public in advance of the people who uttered them had transformed political discourse.[10]

Jefferson Davis's plight underscored another important aspect of the new framework for political debate that rapid means of communication effected. Misquotations and misrepresentations aside, Davis in effect confessed that he spoke in a different voice in Massachusetts than in Mississippi. In fact, this was a long-established, if seldom acknowledged, political practice.

The change in the speed and scope of political reporting accelerated escalation of the sectional issue. For decades, the two-party system had succeeded in containing the sectional conflict; the major political divisions transcended sections. Each of the major national parties was composed of an uneasy alliance of a northern branch and a southern branch. In this way, inter-party rivalry did not focus on territorial differences but on other national issues—such as banking, tariffs, and internal improvements—on which the parties disagreed. To sustain the fragile intra-party coalitions of North and South, the parties and their candidates for elective national offices spoke in different voices on sectional issues in either part of the country.

As political debate became more open, and the state of political reporting more advanced, it became exceedingly difficult to harmonize the interests of both sections in a single party. Ultimately, rapid and intensive circulation of political statements nationwide made it impossible for the parties and candidates to say different things in different areas. And gradually, throughout the 1850's, the political map was redrawn by the emergence of political coalitions whose party lines corresponded to sectional boundaries.[11]

Telegraphic news reporting pressured politicians and parties into taking unambiguous, coherent stands. Statements just spoken were circulated throughout the entire country by the wire service, and over the national system of telegraphs. They were still fresh in the minds of listeners and could not easily be disclaimed. As the *New York Times* put it:

> By the power of the telegraph . . . the public utterances of public men in the furthest sections of the Union are . . . subjected to the criticism of the great centers of population and political activity in all their details, before the leaven of the truth or falsehood which they may contain has time to work in the region for which they were especially designed . . . The telegraph gives the speaker in the furthest East or West an audience as wide as the Union. He is talking to all America . . . immediately, and literally with the emphasis of lightning.[12]

The call by the *New York Times* for presidential candidates who can "utter definite principles" when they "speak to the nation through the magnetic nerves of the telegraph" was ultimately answered by the Republicans, representing the first successful attempt to form a sectional party.

In the opening sentence of his seminal work *The Search for Order,* historian Robert Wiebe observed that during the nineteenth century, America was a society of "island communities." "Weak communication," Wiebe argued, "severely restricted the interaction among these islands and dispersed the power to form opinion and enact public policy."[13] It was early in the second half of the nineteenth century that the telegraph and wire service, through the newspapers, began connecting these islands by means of a common flow of instantaneous information. Once the Civil War was over and Reconstruction was under way, the stream of common information increased considerably.

In 1866, Western Union emerged as a national telegraph monopoly, and the wire service was restructured as a coalition of regional associations led by the New York and the Western Associated Press. Utilizing the superior facilities of Western Union, the regional associations sent their news to New York, then supplied the nation's press with a consolidated report, steadily increasing in scope. Western Union was a harbinger of private-sector national monopolies. It was also an important instrument in facilitating the establishment of institutions of national scale in other fields of business. The emergence of a national news wire service was a proper prelude, and perhaps a useful instrument, to condition Americans to the new world of gigantic, impersonal, non-local institutions to which they had to accommodate as the century progressed.

The Associated Press replaced the great party organs and popular metropolitan newspapers as the primary source of national news. Unlike predecessors such as the Washington *National Intelligencer* or the New York *Weekly Tribune*, the wire service was not a local institution. Wire service reports were a compilation of dispatches, by hundreds of anonymous local agents, printed verbatim in all the country's major newspapers simultaneously. And thus they were not perceived as representing any local institution or interest.

Reporters for the wire service were instructed to send "bare matter of fact." No coloring, no personal references or opinions were permitted.[14] And the concise, matter-of-fact language of the regional agents was hardly altered in the editing process. In fact, editing was usually limited to selection of items. This mode of news presentation, so variant from the standards of individual newspaper correspondents, was shaped by the early managers of the Associated Press. They were not journalists but news vendors, and to them news represented merely "a salable commodity," no different from "a string of onions."

Individual editors presented news items within a preconceived context, reflecting their opinions and personal dispositions, and occasionally their idiosyncrasies. To readers, news in the *New York Tribune* could be identified, to an extent, with the preferences of the "philosopher in the white coat," Horace Greeley; news in the *Herald* with the sardonic, perhaps misanthropic James Gordon Bennett, Sr. In contrast, until Daniel Craig retired from the management of the New York Associated Press after some twenty years of service, Americans had no inkling that an "American Reuter" had ever existed.[15]

The Associated Press monopolized telegraphic news gathering and news distribution in America in the second half of the nineteenth century. Its structure as a national institution—impersonal, non-local, unselfconscious, and hidden—gave wire service news, however partisan, the appearance of objectivity. The Associated Press helped Americans accommodate to a common information environment. By giving news that impressed the minds of Americans a national orientation, it fostered the integration of American society.

Selected Primary Sources

Notes

Acknowledgments

Index

The New York Public Library, Rare Books and Manuscripts Division, Astor, Lenox, and Tilden Foundations, has given its kind permission to use the following material: James Gordon Bennett Papers; Cyrus West Field Papers; Ford Autograph Collection; George Jones Papers; Donald Monroe McNicol Papers; Charles Nordhoff Letters; Henry J. Raymond Letters; Francis O. J. Smith Papers; Western Union Telegraph Company: Records.

Selected Primary Sources

Manuscript Collections

Hiram Alden Papers. Maine Historical Society, Portland, Maine.
David Homer Bates Collection. Library of Congress, Rare Books and Manuscripts Division, Washington, D.C.
Beach Family Papers. New York Historical Society, New York, N.Y.
Moses S. Beach Papers. In the possession of Mr. Brewster Beach, New York, N.Y.
James Gordon Bennett Papers. New York Public Library, New York, N.Y.
John D. Caton Papers. Library of Congress, Washington, D.C.
William W. Clapp Papers. Baker Library, Harvard University, Cambridge, Mass.
James Comly Papers. Ohio Historical Society, Columbus, Ohio.
Cooper Hewitt and Co. Papers. Library of Congress, Washington, D.C.
Charles A. Dana Papers. Library of Congress, Washington, D.C.
Cyrus Field Papers. New York Public Library, New York, N.Y.
Worthington C. Ford Autograph Collection. New York Public Library, New York, N.Y.
Benjamin B. French Papers. Library of Congress, Washington, D.C.
Sidney Howard Gay Papers. Columbia University, New York, N.Y.
Horace Greeley Papers. Library of Congress, Washington, D.C.
Norvin Green Papers. University of Kentucky, Louisville, Ky. (microfilm).
Murat Halstead Papers. Cincinnati Historical Society, Cincinnati, Ohio.
Rutherford B. Hayes Papers. Hayes Memorial Library, Fremont, Ohio (microfilm).
W. R. Holloway Papers. Indiana Historical Society, Indianapolis, Ind.
Edwin M. Hood Papers. Library of Congress, Washington, D.C.
Joseph Howe Papers. Houghton Library, Harvard University, Cambridge, Mass., and National Archives of Canada, Ottawa (microfilm).
George Jones Papers. New York Public Library, New York, N.Y.

John M. Keating Papers. Library of Congress, Washington, D.C.
Amos Kendall Papers. Library of Congress, Washington, D.C.
Victor Lawson Papers. Newberry Library, Chicago, Ill.
Marshall Lefferts Papers. New York Historical Society, New York, N.Y.
Abraham Lincoln Papers. Library of Congress, Washington, D.C.
Manton Marble Papers. Library of Congress, Washington, D.C.
Donald Monroe McNicol Telegraph Collection. New York Public Library, New York, N.Y.
Joseph Medill Papers. Chicago Historical Society, Chicago, Ill.
Samuel F. B. Morse Papers. Library of Congress, Washington, D.C.
Charles Nordhoff Letters. New York Public Library, New York, N.Y.
Henry O'Rielly Papers, Henry O'Rielly Documents, and Henry O'Rielly Telegraph Collection. New York Historical Society, New York, N.Y.
Henry J. Raymond Papers. New York Public Library, New York, N.Y.
Whitelaw Reid Papers. Library of Congress, Washington, D.C. (microfilm).
Rosewater Family Papers. American Jewish Archives, Cincinnati, Ohio (microfilm).
Delavan Smith Papers. William Henry Smith Memorial Library, Indiana Historical Society, Indianapolis, Ind.
Francis O. J. Smith Papers. Maine Historical Society, Portland, Maine, and New York Public Library, New York, N.Y.
William Henry Smith Papers. William Henry Smith Memorial Library, Indiana Historical Society, Indianapolis, Ind., and Ohio Historical Society, Columbus, Ohio.
Edwin M. Stanton Papers. Library of Congress, Washington, D.C.
Melville E. Stone Papers. Newberry Library, Chicago, Ill.
Alfred Vail Papers. Smithsonian Institution Archives, Washington, D.C.
Henry Villard Papers. Houghton Library and Baker Library, Harvard University, Cambridge, Mass.
John Russell Young Papers. Library of Congress, Washington, D.C.

Corporate Records

Associated Press. Corporate Secretary's Office, Associated Press, New York, N.Y.
R. G. Dun and Co. Collection. Baker Library, Harvard University Graduate School of Business Administration, Cambridge, Mass.
Gold and Stock Telegraph Company. Corporate Secretary's Office, Western Union Corporation, Upper Saddle River, N.J., and Electricity Division, Smithsonian Institution, Washington, D.C.
New England Associated Press. Baker Library, Harvard University Graduate School of Business Administration, Cambridge, Mass.

Western Union Telegraph Company. Corporate Secretary's Office, Western Union Corporation; Electricity Division, Smithsonian Institution, Washington, D.C.; and New York Public Library, New York, N.Y. (manuscript division).

Unpublished Public Records

Montreal Board of Trade. National Archives of Canada, Ottawa.
New Brunswick Transportation and Communication. Provincial Archives of New Brunswick, Fredericton, NB (microfilm).
Nova Scotia Telegraph Company. Public Archives of Nova Scotia, Halifax, NS.
Records of the Office of the Secretary of War, RG 107, National Archives, Washington, D.C.
Superior Court of New York. Office of the County Clerk, New York, N.Y.
U.S. Congress. House Judiciary Committee. Unpublished Hearings, RG 233, National Archives, Washington, D.C.

Published Government Documents

Massachusetts. *Legislative Documents* (1859).
New Brunswick. House of Assembly. *Journal* (1847).
New York. *Laws* (1848).
Nova Scotia. House of Assembly. *Journal* (1847–1860).

United Kingdom. Parliament. House of Commons. *Report from the Select Committee on Newspaper Stamps.* London: H. M. Stationery Office, 1851.
U.S. Congress. House. *Telegraphs for the United States.* H. Doc. 15, 25th Cong., 2d sess., 1837.
—— *Electro-Magnetic Telegraph.* H. Ex. Doc. 270, 28th Cong., 1st sess., 1844.
—— *Telegraph Censorship.* H. Rept. 64, 37th Cong., 2d sess., 1862.
—— *Report from the Committee on the Post Office and Post Roads.* H. Rept. 32, 40th Cong., 3d sess., 1869.
—— *Resolutions of the National Typographical Union.* S. Mis. Doc. 13, 41st Cong., 2d sess., 1869.
—— *Postal Telegraph in the United States.* H. Rept. 114, 41st Cong., 2d sess., 1870.
—— *To Connect the Telegraph with the Postal Service.* H. Rept. 6, 42d Cong., 3d sess., 1872.
—— *Telegraph Lines.* H. Rept. 125, 43d Cong., 2d sess., 1875.
U.S. Congress. Senate. *Letter from the Postmaster General Relative to the Suppression of Private Expresses, by the Establishment of a Line of Telegraphic*

Communication between Mobile and New Orleans. S. Ex. Doc. 373, 29th Cong., 1st sess., 1846.

——*Memorial of the Western Union Telegraph Company.* S. Misc. Doc. 86, 42d Cong., 2d sess., 1872.

——*Connecting the Telegraph with the Postal Service.* S. Rept. 242, 42d Cong., 3d sess., 1872.

——*The Postal Telegraph.* S. Rept. 242, 43d Cong., 1st sess., 1874.

——*Telegraph Lines as Post Roads.* S. Rept. 624, 43d Cong., 2d sess., 1875.

——*Competing Telegraph Lines.* S. Rept. 805, 45th Cong., 3d sess., 1879.

——*Report of the Committee on Post Offices and Post Roads on Postal Telegraph.* S. Rept. 577, 48th Cong., 1st sess., 1884.

——Committee on Education and Labor. *Report of the Committee of the Senate upon the Relations between Labor and Capital, and Testimony Taken by the Committee.* 48th Congress (entire), 1883–1885.

Legal Cases

State of Massachusetts. Superior Court, Suffolk County. *Commonwealth by Indictment vs. Francis O. J. Smith. Report of Testimony at the Trial in the Superior Court for Criminal Business for Suffolk County.* May 1865.

State of Massachusetts. Superior Court, Suffolk County. *Commonwealth v. Francis O. J. Smith,* 2 Allen 93 (Mass. 1865).

State of Massachusetts. Superior Court, Suffolk County. *Commonwealth vs. F. O. J. Smith, Opening Arguments of the Defendant.* February 1866.

Commonwealth of Massachusetts vs. Francis O. J. Smith on Indictment Procured by the Conspiracy and Perjuries of Daniel H. Craig and His Witnesses. Portland, Maine, 1867.

State of Massachusetts. Superior Court, Suffolk County. *Commonwealth vs. F. O. J. Smith. Deft's Argument against the Power of the District Attorney to Enter a "Nolle Prosequi."* January 1867.

Opening and Closing Arguments of Hon. Thomas Hayes and Hon. Francis O. J. Smith before the Superior Court of Suffolk County, Mass., on a Motion for a Retrial in the Case of the State vs. Smith. Portland, Maine, 1868.

Opinion of the Supreme Court of the State of New York in the Case of John J. Kiernan vs. the Manhattan Quotation Telegraph Company and Francis A. Abbot. New York: Douglas Taylor, Printer, 1876.

Inter-Ocean Publishing Co. v. Associated Press, 184 Ill. 438, 56 N.E. 822, 825 (1900).

The reports of the proceedings in the case of *Commonwealth v. Smith* may be found in the Smith Papers, Maine Historical Society, Portland, Maine, and in the library of Harvard Law School, Cambridge, Mass.

Pamphlets, Broadsides, and Circulars

Adams, F. C. *The Two Smiths.* 1880.

The American Newspaper Press vs. the Threatened Telegraph Monopoly of F. O. J. Smith and Kendal [sic] *Controllers of Morse's Plan.* 1848.

American Telegraph Company. *Answer and Remonstrance of the American Telegraph Company to the Memorial of the Magnetic and New England Union Telegraph Companies.* 20 April 1858.

—— *Reply of the Executive Committee to the Pamphlet of D. H. Craig.* New York, 1860.

Anderson, John A. *A Postal Telegraph Essential to the Freedom of the American Press and the Prosperity of the American People.* Washington, 1 March 1884.

Articles of Association and Charter from the State of Maryland of the Magnetic Telegraph Company, Together with the Office Regulations and the Minutes of the Meetings of Stockholders and Board of Directors. 1845–1850.

Articles of Association of the New York and Boston Magnetic Telegraph Association, Together with the Office Regulations, Records of the Meetings of the Trustees, Stockholders, and Directors. New York: Chatterton and Christ, 1848.

Atlantic Cable Mismanagement: Correspondence between J. W. Simonton and Cyrus Field and Others. New York: Union Printing House, 1871.

Atlantic Lake and Mississippi Telegraph. *A Brief Sketch of the Origin and Progress of the Telegraph Lines under the Management of Henry O'Rielly.* October 1848.

—— *Equal Rights and No Monopoly in Telegraphy.* 1848.

—— *New Telegraphs, Reduced Rates, etc.* 1848.

—— *New York and New England Telegraph.* 1848.

—— *Report of the Committee Appointed by the Citizens of Louisville to Examine the Subject of the Telegraph and the Responses of the Chairman of the Committee* [George D. Prentice] *to the Hon. Amos Kendall's Attack upon the Committee.* 1847.

Barnum, Zenus, Hiram O. Alden, and Robert W. Russell. *The American Telegraph Company and the New York Associated Press.* New York, 22 June 1860.

Circular from the Executive Committee of the Western Associated Press. Cincinnati, 15 April 1867.

Copp, Joseph A. *A Discourse Preached in the Broadway Church, Chelsea* [Mass.]. Boston: T. R. Martin and Son, 1858.

Craig, Daniel H. *Annual Report of the General Agent.* New York, 1 January 1862.

—— *Answer of Daniel H. Craig to the Interrogatories of the United States Senate Committee on Education and Labor.* New York, 1883.

—— *The Convicted Libeller.* New York, 1863.

—— *To Editors.* New York, 9 November 1866.

—— *Letter to Freeman M. Edson.* New York, 1853.

—— *Machine Telegraphy To-Day.* New York, September 1891.

—— *To the Members of the New York Associated Press.* New York, November 1866.

—— *New York Merchants Exchange and News Association.* Broadside. New York, 17 January 1867.

—— *Press News Reports and Letter Correspondence by Machine Telegraphy.* New York, 1888.

—— *A Reply to the Falsehoods of the Executive Committee.* New York, 7 July 1860.

—— *A Review of an Exposition of the Differences Existing between Different Presses and Different Lines of Telegraph, Respecting the Transmission of Foreign News, Being A Letter and Accompanying Documents, Addressed to the Government Commissioners of the Nova Scotia Telegraph.* Halifax, 1950.

—— *The Telegraph and the Press: A Report to the Press of the Actions of Various Committees and Resolutions Leading up to an Increase in Tariff on Press Messages.* 1860.

Debate in the Massachusetts Senate on the Bill to Incorporate the Boston and Halifax Telegraph Cable Co., 24 and 26 March 1859. Boston: Henry W. Dutton and Son, 1859.

General News Association of the City of New York. *Rules.* 1870, 1874.

Gold and Stock Telegraph Company. *Universal Stock Printer.* 1872.

Howe, Joseph. *To the Electors of the Province of Nova Scotia.* Broadside. Amherst, 15 August 1851.

—— *Information for the People: To the Freeholders of the County of Halifax.* Broadside. Halifax, 30 October 1850.

Kendall, Amos. *To the Stockholders of the American Telegraph Company.* 27 August 1860.

Kendall, Amos, William M. Swain, Zenus Barnum, H. O. Alden, and R. W. Russell. *The Telegraph and the Press.* 18 April 1860.

Kennedy, Grammond. *The Venezuelan "Investigation."* 1876.

Lefferts, Marshall. *Address to the Stockholders of the New York and New England Telegraph Company.* New York: Snowden, January 1851.

Magnetic Telegraph Company and Union Telegraph Company. *British Telegraph Monopoly.* 1858.

—— *Memorial of the Magnetic Telegraph Company and the New England Union Telegraph Company to the Congress of the United States.* 10 March 1858.

—— *Memorial to the Senate and House of Representatives of the United States of the Magnetic and New England Union Companies, in Reply to the Remonstrance of the American Telegraph Company.* 1858.

—— *Regulations of the Magnetic Telegraph Company.*

National Typographical Union. *Proceedings of the Seventeenth Annual Session.* New York, 1869.

New York Associated Press. *Annual Report.* 1862, 1872 [flimsy], 1876.

New York Daily Graphic. *The Press and the Telegraph: The Ramifications and Oppressions of Two Gigantic Monopolies.* New York, 1874.

New York and New England Telegraph Company. [Annual] *Report.* 31 March 1852.

New York and Sandy Hook Telegraph Company. *Articles of Association and Incorporation.* New York, 1852.

North American Telegraph Association. *Proceedings.* 1858–1860.

Ohio and Mississippi Telegraph Company. *Rules for the Government of Offices.* St. Louis, 1848.

One of the Reasons for Telegraphic Reform: Power and Tyranny of the Associated Press—the Character of Its Manager, James W. Simonton. 1873.

O'Rielly, Henry. *To the Citizens of Cincinnati.* 1847.

—— *To the Citizens of St. Louis.* 1847.

Orton, William. *Government Telegraphs: Argument of William Orton, President of the Western Union Telegraph Company on the Bill to Establish Postal Telegraph Lines, Delivered before the Select Committee of the United States House.* New York, 1870.

Phillips, Walter Polk. *A Mild Corrective.* New York: Evening Post, 1883.

Press Association of the Confederate States of America. *First Annual Meeting, Minutes of the Board of Directors, and Quarterly Report of the Superintendent.* Montgomery, Alabama [?]: Memphis Appeal Job Printing Establishment, 1864.

—— *Minutes, Rules, and Report of the Superintendent.* Griffin, Georgia: Hill and Swayze's Printing House, 1863.

Proceedings of the American Telegraph Convention, Composed of Delegates of the Principal Morse Telegraph Companies in the United States. Philadelphia, 1853.

Russell, Robert W. *Remarks of R. W. Russell, One of the Executive Committee of the American Telegraph Company, in Reply to the Statements of Messrs. Abram S. Hewitt, Cyrus W. Field, Henry J. Raymond, and Others, Made at the Meeting of Stockholders, on June 29th, 1860.* New York: William C. Bryant and Co., 1860.

—— *Statistics, etc., of the American Telegraph Company.* New York, 20 August 1860.

Simpson, William H. *An Address Delivered before the Maine Editors and Publishers Association.* Portland, Maine: B. Thurston and Co., 1876.

Smith, Francis O. J. *An Exposition of the Differences Existing between Different Presses and Different Lines of Telegraph, Respecting the Transmission of Foreign News, Being a Letter and Accompanying Documents, Addressed to the Government Commissioners of the Nova Scotia Telegraph.* Boston, 1850.

—— *Mr. Craig's "Superior Activity" in Telegraphing.* Portland, Maine, 1851.

Smith, William Henry. *Government Regulation of the Press: Testimony Taken before the Senate Committee on Post Offices and Post Roads.* New York, 1884.

—— *The Public Press as the Advocate of Human Rights and the Champion of the Interests of the Common People.* 1893.

Somerville, W. B. *The Newspaper Service of the Western Union Telegraph Company.* December 1883.

Southern Press Association. *Proceedings.* 1869.

Speech of Judge Johnson at Delaware, Ohio, Being a Brief Review of the Claims of S. J. Tilden as Set Forth by the Cincinnati Enquirer before His Nomination. Cincinnati: Robert Clarke and Co., 1876.

Sumner, Charles A. *The Political Retrospect and Our Resolve and Duty.* San Francisco: Bacon and Company, 1877.

—— *Speech of Charles Sumner of California in the House of Representatives.* Washington, 1884.

—— *Speech of Hon. Charles A. Sumner of California in the House of Representatives, February 28, 1885.* Washington, 1885.

Wells, David A. *The Relation of the Government to the Telegraph.* New York, 1873.

Western Associated Press. *Proceedings.* 1867–1897.

Western Union Telegraph Company. *Annual Report.* 1867–1888.

—— "History of the Magnetic Telegraph," in *The Western Union Telegraph Company: Its Past, Present, and Future.* New York, 1883.

—— *The Proposed Union of the Telegraph and Postal System: Statement of the Western Union Telegraph Company.* Cambridge, Mass., 1869.

Periodicals

American Telegraph Magazine (1852–1853)
Congressional Globe
Congressional Record
Journal of the Telegraph (1867–1884)
Journalist (1884–1885)
National Telegraph Review (1852)
Shaffner's Telegraph Companion (1854–1855)
Telegrapher (1864–1877)

Notes

Introduction

1. Undated clipping from the *San Francisco Post,* William Henry Smith papers, Ohio Historical Society, Columbus, Ohio.
2. Edmund Burke, quoted in George S. Merriam, *The Life and Times of Samuel Bowles* (New York: The Century Co., 1885), p. 363.
3. James W. Carey, "Canadian Communication Theory: Extensions and Interpretations of Harold Innis," in *Studies in Canadian Communications,* ed. Gertrude Joch Robinson and Donald F. Theal (Montreal, 1975), pp. 38–45.
4. Among Innis's major published works are *Empire and Communications* (Oxford: Clarendon Press, 1950); *The Bias of Communication* (Toronto: University of Toronto Press, 1951); and *Changing Concepts of Time* (Toronto: University of Toronto Press, 1952). A typewritten manuscript of Innis's notes, entitled "History of Communication," is available on microfilm. In these fragments, there is strong emphasis on the history of the press.
5. Alexis de Tocqueville, *Democracy in America,* ed. Phillips Bradley (New York: Vintage Books, 1954), vol. 2, pp. 119–122. Whether de Tocqueville was referring to the number of newspaper establishments, newspaper copies, or both is somewhat moot.
6. Ben: Perley Poore, *Perley's Reminiscences of Sixty Years in the National Metropolis,* 2 vols. (Philadelphia: Hubbard Brothers, 1886); Frederick B. Marbut, *News from the Capital: The Story of Washington Reporting* (Carbondale: Southern Illinois University Press, 1971); Marbut, "Decline of the Official Press in Washington," *Journalism Quarterly* 33 (Summer 1956): 335–347; Culver H. Smith, *The Press, Politics, and Patronage: The American Government's Use of Newspapers, 1789–1875* (Athens: University of Georgia Press, 1977); William E. Ames, *A History of the National Intelligencer* (Chapel Hill: University of North Carolina

Press, 1972); Thomas C. Leonard, *The Power of the Press: The Birth of American Political Reporting* (New York: Oxford University Press, 1986); Donald Ritchie, *Press Gallery: Congress and the Washington Correspondents* (Cambridge, Mass.: Harvard University Press, 1991).

7. Clarence Darrow, quoted in Innis, *Bias of Communication*, p. 187.
8. Innis, *Bias of Communication*, p. 139. Innis stated similarly on p. 187 that "freedom of the press had been an essential ingredient of the monopoly for it obscured monopolistic characteristics."
9. For other broad statements of the significance of the telegraph, see Daniel J. Czitrom, *Media and the American Mind: From Morse to McLuhan* (Chapel Hill: University of North Carolina Press, 1982); James W. Carey, "Technology and Ideology: The Case of the Telegraph," *Prospects* 8 (1983): 303–325; James W. Carey and John J. Quirk, "The Mythos of the Electronic Revolution," *American Scholar* 39 (Spring 1970): 219–241 (Summer 1970): 395–424. Versions of these last two articles have appeared more recently in James W. Carey, *Communication as Culture: Essays on Media and Society* (New York: Routledge, 1992).
10. Innis, *Bias of Communication*, p. 169.
11. Ibid., pp. 59, 170. Allan Pred, who studied spatial biases in the circulation of information in the telegraphic period, found that "major centers compounded their information availability and circulation advantages during the period." Whereas Pred points out circumstances that diminished the telegraph's potential to level information advantages, he appears to downplay the fact that time lags in receipt of economic information in the hinterland were, in fact, dramatically reduced. Allan Pred, *Urban Growth and City-Systems in the United States, 1840–1860* (Cambridge, Mass.: Harvard University Press, 1980), pp. 142–165.
12. Marshall McLuhan, *Understanding Media: The Extensions of Man* (New York: McGraw-Hill, 1965), pp. 255–256. According to McLuhan, this effect was produced by the mere existence of the telegraph, its capacity for interaction in real time serving as a metaphor that shaped consciousness. Moreover, although McLuhan viewed telegraphic technology as interactive, he also attributed the effect of the telegraph on improving human contact across distances to the simultaneity of information, which promoted a sense of shared experience. In this respect, his argument is convincing.
13. The telegraph was initially promoted and employed as an instantaneous two-way medium; see, e.g., *New York Herald*, 4 June 1844, 6 December 1847; *National Police Gazette*, 30 May 1846; *Philadelphia Dollar Newspaper*, 1 January 1846; *Albany Evening Atlas*, 6 July 1846; Edward Lind Morse, *Samuel F. B. Morse: His Letters and Journals* (Boston:

Houghton Mifflin Co., 1914), vol. 2, p. 224. Morse's assistants publicized this feature by arranging telegraphic contests between the chess clubs of Washington and Baltimore. From early in the process of its diffusion, however, the telegraph functioned as a one-way medium par excellence. This argument is developed in my paper "When Bad Things Happen to Good Technologies: Telegraphic Diffusion in Nineteenth-Century America" (delivered at the International Symposium "Technological Pessimism, Modern Societies, and Their Environments," Tel Aviv, 1992).

14. McLuhan, *Understanding Media*, p. 256.
15. Alfred D. Chandler, Jr., *The Visible Hand: The Managerial Revolution in American Business* (Cambridge, Mass.: Harvard University Press, 1977), pp. 81–82.
16. I use "Associated Press" to denote both the original association of New York dailies that became commonly known as the "New York Associated Press" (and after 1856, officially styled the "General News Association of the City of New York") and the combination of the New York Associated Press and its affiliated regional press associations after 1866.
17. Victor Rosewater's work was published in New York by D. Appleton and Co. Schwarzlose's work *The Nation's Newsbrokers* consists of two volumes: vol. 1, *The Formative Years: From Pretelegraph to 1865;* and vol. 2, *The Rush to Institution: From 1865 to 1920* (Evanston, Ill.: Northwestern University Press, 1989–1990).

1. "The News Newer"

1. "The News Newer" is from the *Boston News-Letter,* 10 August 1719. In this issue, the proprietor of the *News-Letter* called for support for more frequent publication, stating that "it was impossible with half a Sheet a Week to carry on all the Publick Occurrences of Europe, with those of this, our Neighboring Provinces, and the West Indies. To make up which Deficiency, and the News Newer and more acceptable, he has since Printed every other week a Sheet, whereby that which seem'd Old in the former half Sheets, becomes New." The quotation on speed is from U.S. Congress, House, *Telegraphs for the United States,* H. Doc. 15, 25th Cong., 2d sess., 1837, p. 30.
2. One notable exception to the dearth of studies of this aspect is Stephen Kern, *The Culture of Time and Space, 1880–1918* (Cambridge, Mass.: Harvard University Press, 1983). The impact of technologies that Kern viewed as shaping the culture of the late nineteenth and early twentieth century appear fully congruent with the impact of the telegraph a

half century earlier. See also Wolfgang Schivelbusch, *The Railway Journey* (New York: Urizen, 1979).

3. Allan Pred, *Urban Growth and the Circulation of Information: The United States System of Cities, 1790–1840* (Cambridge, Mass.: Harvard University Press, 1973), p. 38. In 1817 the time-distance consisted of 2.9 days between Washington and Philadelphia; 1.1 days between Philadelphia and New York; and 2.2 days between New York and Boston (pp. 46, 54). Pred considers 2.8 days the average time-distance between Boston and Washington in 1841 (p. 54). There is, however, extensive evidence (which Pred acknowledges) that by the mid-1840's news items were being transmitted between the cities in less than twenty-four hours by means of regular and special expresses.
4. The concept of time-distance is best developed in John C. Lowe and S. Moryadas, *The Geography of Movement* (Boston: Houghton Mifflin Co., 1975).
5. It is important to note that only the signal was virtually instantaneous. Because the Morse system required a rewriting of the message in code, the significance of the economy in time of transmission declined in proportion to the length of the message. For example, the transmission of the Old Testament between Boston and New York would have taken less than ten hours by express, but more than five days by telegraph. (This calculation is based on Alfred Vail's estimates of speed of telegraphic transmission in his memorandum to Post Master General Cave Johnson, 26 November 1846. Vail papers, Smithsonian Institution Archives, Washington, D.C.)
6. Quoted in: A Journalist [Isaac C. Pray], *Memoirs of James Gordon Bennett and His Times* (New York: Stringer and Townsend, 1855), p. 194.
7. Richard B. Eide, "The Influence of Editorship and Other Forces on the Growth of the *St. Paul Pioneer-Press,* 1849–1909" (M.A. thesis, University of Missouri, 1939), p. 77. See also pp. 100–103 for problems of news scarcity.
8. Charles T. Congdon, *Reminiscences of a Journalist* (Boston: James R. Osgood and Company, 1880), p. 22.
9. Ibid., p. 254. Samuel Bowles editorialized in the *Springfield Republican* of 4 January 1851 that "nothing can be more evident to the public, and nothing certainly is more evident to publishers of newspapers, than that there is a great deal more news nowadays than there used to be . . . Publishers of country weeklies used to fish with considerable anxiety in a shallow sea, for matter sufficient to fill their sheets, while dailies only dreamed of an existence . . . Now all is changed. The increase of facilities for the transmission of news brought in a new era." Quoted in Merriam, *Samuel Bowles,* p. 98. See also Judge Kent's description of the *Albany Gazette* in the 1820's in a lecture before the

Young Men's Association of Albany in 1854, quoted in Frederic Hudson, *Journalism in the United States from 1690 to 1872* (New York: Harper and Brothers, 1873), p. 280.

10. Thurlow Weed, in the *Evening Journal* of 1852, quoted in Hudson, *Journalism in America,* p. 226. William T. Cogeshall, *The Newspaper Record* (Philadelphia, 1856), p. 163. And see James L. Crouthamel, "The Newspaper Revolution in New York, 1830–1860," *New York History* 45 (April 1964): 92–93.
11. On the relation between the volume of social interaction and the media, see Samuel N. Eisenstadt, *Propaganda in Modern Society* (Jerusalem, 1961), pp. 7–25 (in Hebrew).
12. The "no news" announcements were frequently expressions of lack of interest in news received by the exchanges, not only reflections of their non-receipt. See, e.g., Thomas H. Baker, *The Memphis Commercial Appeal* (Baton Rouge: Louisiana State University Press, 1971), p. 35; compare Richard B. Kielbowicz, "News in the Mails, 1690–1863: The Technology, Policy, and Politics of a Communication Channel" (Ph.D. diss., University of Minnesota, 1984), p. 346. The general perception of pre-modern American society as localized and inwardly oriented emerges from studies emphasizing "modernization." Robert H. Wiebe, *The Search for Order, 1877–1920* (New York: Hill and Wang, 1967), is a seminal expression of this approach, expanded more recently in *The Opening of American Society* (New York: Vintage Books, 1984). See also Richard D. Brown, *Modernization: The Transformation of American Life, 1600–1865* (New York: Hill and Wang, 1976); Brown, "Modernization and the Modern Personality in Early America, 1600–1865: A Sketch of a Synthesis," *Journal of Interdisciplinary History* 2 (December 1972): 201–228. Samuel N. Eisenstadt, "Studies of Modernization and Sociological Theory," *History and Theory* 13 (1974): 225–252, is a good overview of modernization theory. See also Dean C. Tipps, "Modernization Theory and the Comparative Study of Societies: A Critical Perspective," *Comparative Studies in Society and History* 15 (March 1973): 199–226; Joyce Appleby, "Modernization Theory and the Formation of Modern Social Theories in England and America," *Comparative Studies in Society and History* 20 (April 1978): 259–285; James Tagg, "Agency, Autonomy, and Authority: The Ethos and Modernization," *Canadian Review of American Studies* 15 (Winter 1984): 451–463.
13. The standard summary of the development of local news is David J. Russo, "The Origins of Local News in the U.S. Country Press, 1840's–1870's," *Journalism Monographs* 65 (February 1980). See also Donald L. Shaw, "At the Crossroads: Change and Continuity in American Press News, 1820–1860," *Journalism History* 8 (Summer 1981): 38–50;

Cogeshall, *Newspaper Record*, p. 164; Joseph E. Chamberlin, *The Boston Transcript* (Cambridge: Houghton Mifflin Co., 1930), pp. 20, 82; Kenneth Q. Jennings, "Political and Social Force of the New Jersey Press Association, 1857–1939" (M.A. thesis, Columbia University, 1941), p. 3.

14. Leonard, *Power of the Press*, p. 66.
15. Ibid., p. 69.
16. Developments in the frequency and speed of European arrivals are summarized in George R. Taylor, *The Transportation Revolution: 1815–1860* (New York: Holt, Rinehart, and Winston, 1962), pp. 144–148, and detailed in Robert G. Albion, *Square-Riggers on Schedule* (Princeton: Princeton University Press, 1938), especially pp. 175–201, 274, and 317–325; Robert G. Albion, *The Rise of New York Port (1815–1860)* (New York: Charles Scribner's Sons, 1939), pp. 50–55, 394–397. See also Pred, *Urban Growth*, pp. 26–35.
17. In contemporary media usage, news events are categorized as spot news, developing news, or continuing news. Gaye Tuchman has pointed out that rather than characterizing variance in the intrinsic qualities of news events, these designations reflect the different constraints on news organizations in covering occurrences. In fact, most spot news also develops and, to a varying degree, continues. Tuchman, *Making News: A Study in the Construction of Reality* (New York: The Free Press, 1978), pp. 47–63.
18. Hudson, *Journalism in America*, pp. 394–395. Frank Luther Mott, *American Journalism: A History of Newspapers in the United States through 260 Years, 1690 to 1950* (New York: The Macmillan Company, 1950), p. 218, relates the same anecdote. For a general statement of early editors' inactivity, see James Edward Rogers, *The American Newspaper* (Chicago: University of Chicago Press, 1909), p. 66.
19. Thurlow Weed, quoted in Frederick Follet, *History of the Press of Western New-York* (reprint ed., Harrison, N.Y.: Harbor Hill Books, 1973), p. 49.
20. Kielbowicz, "News in the Mails," p. 345.
21. Quotation from a letter of reminiscences written in 1846 by E. J. Fowle, in Follet, *Press of Western New-York*, p. 51.
22. The best accounts of these arrangements are Richard Kielbowicz, *News in the Mail* (Westport, Conn.: Greenwood Press, 1989), and "The Press, Post Office, and Flow of News in the Early Republic," *Journal of the Early Republic* 3 (Fall 1983): 255–280; "News-Gathering by Printers' Exchanges before the Telegraph," *Journalism History* 9 (Summer 1982): 42–48; "Origins of the Second-Class Mail Category and the Business of Policymaking, 1863–1879," *Journalism Monographs* 96 (April 1986); and "Speeding the News by Postal Express, 1825–1861: The Public Policy

of Privileges for the Press," *The Social Science Journal* 22 (January 1985): 49–63.

23. James J. Belcher, "Newspaper Curiosities," *Harper's New Monthly Magazine* 35 (September 1867), p. 489.
24. In 1810, daily newspapers appeared in only six of twenty-two states and territories. Cogeshall, *Newspaper Record,* p. 126. The table in Alfred M. Lee, *The Daily Newspaper in America* (New York: The Macmillan Company, 1947), pp. 718–719, is especially useful for following this aspect of the development of daily newspaper publication.
25. Kielbowicz describes how the news of President Harrison's death reached the *Jefferson Inquirer* (Jefferson City, Missouri) after five stages of transmission, which took eleven days. "News in the Mails," pp. 347–348.
26. Robert G. Albion, "The 'Communication Revolution,'" *American Historical Review* 37 (July 1832): 718–720. The standard account of this phenomenon is Taylor, *The Transportation Revolution.*
27. Leonard, *Power of the Press,* pp. 75–90.
28. For general discussions, see Phyllis Wilson, "The Nature of News," in *Journalism, Communication, and the Law,* ed. G. Stuart Adam (Scarborough, Ontario: Prentice Hall, 1976), pp. 23–33; Bernard Roshco, *Newsmaking* (Chicago: University of Chicago Press, 1975), pp. 9–19; Wilbur Schramm, "The Nature of News," *Journalism Quarterly* 26 (September 1949): 259–269.
29. Shaw, "At the Crossroads," p. 48 (adapted from table 8).
30. Taylor, *Transportation Revolution,* p. 143.
31. Ibid., p. 142.
32. Ibid., p. 143.
33. Ibid., pp. 138–144.
34. Hudson, *Journalism in America,* p. 432, provides an insightful analysis of the relation between the revolution in transportation and in journalism.
35. Statistics of the number of periodicals in general and of dailies appear in numerous primary and secondary works, and they are only slightly divergent. The most recent statistical account is in Allan Pred, *Urban Growth,* p. 21, and includes data up to 1840. These data (with minor corrections) together with figures provided by William A. Dill, *Growth of Newspapers in the United States* (Kansas City: University of Kansas, 1928), p. 31, for subsequent periods, formed the basis for calculating the percentages presented in the text.
36. *New York Journal of Commerce,* 1 September 1828, quoted in Joseph P. Thompson, *Memoir of David Hale* (Hartford: E. Hunt, 1850), p. 48, and see also pp. 46–59, 136–144; Hudson, *Journalism in America,* p. 346; William H. Hallock, *Life of Gerard Hallock* (New York: Oakley, Mason,

and Co., 1869); Frank M. O'Brien, *The Story of the "Sun"* (New York: D. Appleton and Company, 1928), p. 82.

37. Affidavits in the case of *Craig v. Kendall,* Superior Court of New York, Office of the County Clerk, New York, N.Y.; James Parton, *Life of Horace Greeley* (Boston, 1872), p. 208. See Chapter 4.
38. *The Boston Herald and Its History* (Boston, 1878), p. 6; James M. Lee, *History of American Journalism* (Cambridge: Houghton Mifflin Co., 1917), p. 218; Oliver Carlson, *The Man Who Made News: James Gordon Bennett* (New York: Duell, Sloan, and Pearce, 1942), p. 200.
39. Parton, *Horace Greeley,* p. 103.
40. The best summary of this voluminous literature on the penny press is Michael Schudson's authoritative discussion in *Discovering the News: A Social History of American Newspapers* (New York: Basic Books, 1978), pp. 12–60. And see also Dan Schiller, *Objectivity and the News: The Public and the Rise of Commercial Journalism* (Philadelphia: University of Pennsylvania Press, 1981). John C. Nerone, "The Mythology of the Penny Press," *Critical Studies in Mass Communication* 4 (December 1987): 376–404, and critical responses and Nerone's reply (pp. 405–422), are a useful conspectus and update.
41. Stephen Botein, "'Mere Mechanics' and an Open Press: The Business and Political Strategy of Colonial American Printers," *Perspectives in American History* 9 (1975): 127–225; Alexander Saxton, "Problems of Class and Race in the Origins of the Mass Circulating Press," *American Quarterly* 36 (Summer 1984): 211–233; Crouthamel, "The Newspaper Revolution." The one exception to the dominance of printers in the group, James Gordon Bennett, was a shifty journalist before the *Herald* was established. He arrived in the United States a lonely immigrant, penniless, vocationless, and, worst of all, "so terribly cross-eyed that when he looked at me with one eye, he looked out at the city hall with the other." William A. Croffut, *An American Procession,* quoted in Carlson, *The Man Who Made News,* p. 262; and compare Parton, "The New York Herald," *North American Review* (April 1866).
42. The literature on this is vast. See, e.g., Crouthamel, "The Newspaper Revolution"; Elwyn Burns Robinson, "The Dynamics of American Journalism from 1787 to 1865," *Pennsylvania Magazine of History and Biography* 61 (October 1937): 435–445. Ultimately, even the most party-minded of the penny press editors, Horace Greeley, had serious differences with his Whig party. Hudson, *Journalism in America,* p. 550.
43. Frederic Hudson remarked that "it took the editors some time to make up their minds what to say on the important topics of the day. This was part of the discipline of the party press . . . Time was taken too, for reflection and consideration." *Journalism in America,* p. 219.
44. Parton, *Horace Greeley,* pp. 105–106; Hudson, *Journalism in America,*

p. 417; Cogeshall, *Newspaper Record,* pp. 147–148; Mott, *American Journalism,* pp. 218–220.

45. Pray, *Memoirs of Bennett,* pp. 181–182; Cogeshall, *Newspaper Record,* p. 142.
46. But see Nerone, "The Mythology of the Penny Press," pp. 383–385.
47. Pray, *Memoirs of Bennett,* p. 181.
48. First editorial of Nathan Hale as editor of the *Boston Repertory and Daily Advertiser,* 7 April 1814, quoted in Hudson, *Journalism in America,* p. 380. Quotation of Greeley is from his testimony before a Select Committee of the House of Commons on a visit to England. United Kingdom, Parliament, House of Commons, *Report from the Select Committee on Newspaper Stamps* (London: H. M. Stationery Office, 1851). Greeley's testimony was reprinted in part in Parton, *Life of Greeley,* and in Hudson, *Journalism in America.* The quotation in the text appears in Hudson, p. 541; see also p. 548. Subsequent references will also be to Hudson's version.
49. The absolute deadline for supplying the newspaper to local subscribers was 7 A.M. (Greeley, "Testimony," quoted in Hudson, *Journalism in America,* p. 545), and it was necessary to meet firm deadlines for the departure of mails and expresses for delivery to out-of-town readers. Considering the large size of each edition, swift printing was crucial if the newspaper was to supply the readers of the first copies off the presses with the latest news.
50. The *New York Sun,* on 2 June 1846, claimed to have printed 30,000 extras the previous day. Its regular morning edition was of a city-leading 50,000 copies. Frederic Hudson, in *Journalism in America,* p. 421, described the sending of two packages of extras printed by the *Boston Mail* on behalf of the *Herald* to New York. Each package contained several thousand copies. The *New York Express,* 29 May 1844, reported the sale of 5,000 copies of the late edition within 2 hours on the previous day.
51. There was, however, one circumstance that worked in the opposite direction. Under previous conditions, when the newspaper made one daily appearance, the difference between receiving information just prior to or just after the "closing" of the newspaper was all important. The emergence of the extra provided more leeway, since the newspaper could score a beat even after the close of the morning edition.
52. In 1851, the *Tribune* regularly published two evening editions. Greeley, "Testimony," quoted in Hudson, *Journalism in America,* p. 540. The *New York Express* made the publication of numerous editions throughout the day its routine schedule (eventually it even stopped its morning edition). This was probably related to the fact that the *Express* was established as a printed extension of Hudson's News Room and was

modeled to some extent on the continually updated log books of the establishment. Hudson, *Journalism in America,* pp. 518–520.

53. *New York Tribune,* 26 June 1850.
54. *New York Sun,* 14 December 1843, quoted in O'Brien, *Story of the "Sun,"* pp. 96–97. Other means of promotion that the penny press inaugurated were also dependent on timely news, and their effectiveness motivated editors to be the first to get this news. What became known as "the news-crying nuisance" represented the efforts of newsboys to advertise the paper by calling attention to the important news in the issue. (See *First Five Years of Sabbath Reform,* 1862, pp. 7–8, quoted in Lee, *Daily Newspaper,* p. 394.) Street sales also encouraged the novel practice of printing news on the front page, so as to catch the eye of potential customers. Such means of promotion demonstrated that it was now "news which sold the newspaper, not the newspaper which sold news."
55. For a general treatment see Michael Schudson, "Deadlines, Datelines, and History," in *Reading the News,* ed. Robert C. Manoff and Michael Schudson (New York: Pantheon Books, 1986), pp. 79–108.
56. John W. Regan claimed that the *Journal of Commerce* "erected a semaphore on the highlands near Sandy Hook to which they signaled news and this in turn was transmitted to Staten Island." "The Inception of the Associated Press," *Transactions of the Nova Scotia Historical Society* 19 (1912), p. 4; and this was also the version of Lee, *History of American Journalism,* p. 219. Numerous other sources attest to the use of the semaphore by the *Journal* but do not claim that the paper established it.
57. William F. S. Shanks, "How We Get Our News," *Harper's New Monthly Magazine* 34 (May 1867), p. 513.
58. Each innovation in transportation that implied speed was received with great public enthusiasm. The extraordinary coverage of these developments by the press attests to such public fascination. Thus Thurlow Weed represented the press on the first railroad trip in America, and James Gordon Bennett embarked on the first voyage of an Atlantic steamer to Europe. It is in this light that the frequent enthusiastic descriptions of the achievements of the rapid presses, and above all the accomplishments in speedy news gathering, should be considered. Some important data on this aspect of the transportation revolution are presented in Charles M. Pickett, "Six New York Newspapers and Their Response to Technology in the Nineteenth Century" (Ph.D. diss., University of Minnesota, 1959); and see Pickett, "Technology and the New York Press in the Nineteenth Century," *Journalism Quarterly* 37 (Summer 1960): 398–407.

59. Regan, "Inception of the AP," pp. 13–14.
60. Common terminology differentiated between penny papers and six-penny papers. Whereas the former epithet correctly states the price of the new newspapers, the latter is somewhat misleading. The established newspapers were available by subscription, commonly at $8 to $10 per annum, or some 2.5¢ to 3¢ per copy. Six pennies was the price of single issues of established journals when purchased in the newspaper office, from carriers, or as extras on the rare occasions when editors acquired important information between daily editions and published the papers for spot sales. These sale situations were somewhat similar to the way penny papers were vended; hence, they served as the basis for price comparison, or rather for the contrast in price.
61. Hudson, *Journalism in America,* p. 372.
62. Innis, *Bias of Communication,* p. 162. See also Lee, *Daily Newspaper,* pp. 315–321; Schudson, *Discovering the News,* pp. 18–21; and Frank Presbery, *The History and Development of Advertising* (New York: Doubleday, Doran, and Company, 1929).
63. *New York Herald,* 12 July 1848.
64. Horace Greeley, quoted in Hudson, *Journalism in America,* p. 543.
65. The estimate for 1832 is from Cogeshall, *Newspaper Record,* p. 142; and for 1836, from O'Brien, *Story of the "Sun,"* p. 31. The *Philadelphia Public Ledger* claimed, in 1836, that the circulation of the New York penny press alone was 60,000 copies daily, but that figure appears inflated.
66. Hudson, *Journalism in America,* p. 440.
67. Newspaper proprietors, by reinvesting revenues into efforts to speed the news, indicated that they considered this aspect of their enterprise the key to success. Thus Cogeshall noted that it was the policy of the penny press "to gain a large circulation, and to support enterprise in the collection of news by the profits." *The Newspaper Record,* p. 129. James Parton states that "for some years every dollar that the *Herald* made was expended in news." "The New York Herald," p. 396. See also the *New York Herald,* 17 August 1837; Hudson, *Journalism in America,* pp. 421, 446.
68. In 1840, New York dailies circulated 60,000 copies. Simon N. D. North, *The Newspaper and Periodical Press: A Special Report for the Tenth Census* (Washington: Government Printing Office, 1884), p. 98. By 1850 they circulated 153,621 copies. Cogeshall, *Newspaper Record,* p. 144.
69. James Carey, "Canadian Communication Theory," p. 40.
70. Hudson, *Journalism in America,* p. 189.
71. Ibid., p. 334.
72. Dan Schiller, in *Objectivity and the News,* while acknowledging the achievements of the penny press in this field, makes a much greater

issue of its achievements in providing "cheap, value free information" (p. 10), or objectivity, in the sense of freedom from the elite's definitions of social values. Schiller's arguments are based largely on identifying the readership of the penny papers with "the artisans and mechanics of New York" (p. 17)—a disputable proposition. His interpretations of social developments in the period are based closely on work by Sean Wilentz. See, e.g., Sean Wilentz, "On Class and Politics in Jacksonian America," *Reviews in American History* (December 1982): 45–63, and *Chants Democratic: New York City and the Rise of the American Working Class, 1788–1850* (New York: Oxford University Press, 1984). Compare Edward Pessen, "Who Has Power in the Democratic Capitalistic Community? Reflections on Antebellum New York City," *New York History* 58 (April 1977): 129–155.

73. Quoted in Pray, *Memoirs of Bennett,* p. 375.
74. Frederic Hudson described Hudson's news room as occupied by "brokers, bankers, merchants, shipmasters, and gossips." *Journalism in America,* p. 517. Proprietors of news rooms permitted the press to subscribe and were not concerned about losing the business of the mercantile community, since subscribers to the news room still held an advantage over the newspaper-reading public: they could immediately act upon the information in the marketplace before an account appeared in a newspaper edition.
75. Hudson, *Journalism in America,* pp. 364–365, 391.
76. *New York Herald,* 26 January 1847.
77. Newspapers gained the lead in the release of timely news in the periphery, as well as in urban centers, owing to the advantages of centralized distribution. Large-scale delivery to towns in the hinterland made for economies in using express services. Thus the directors of one of the early railroads stated in 1834 that "the company now sends an express daily from one commercial city to another . . . The daily papers of this city are sent by this conveyance, but merchants' letters, of the utmost importance to them in business, are not less than two days going under contract." Quoted in Hudson, *Journalism in America,* p. 432.

The outsider status of the penny press editors gave them a perceived (perhaps even real) advantage in the reliability of their commercial reports. Thus, as early as 23 July 1835, Bennett stated in an editorial that the *Herald* was "the only paper in the city which gives authentic and correct daily reports of Wall Street operations, stocks, and the money market." On 20 February 1869, Bennett, in a retrospective description of his early "money articles," stated that in the hands of personally interested financial commentators "the financial report becomes a dangerous machine; but it must not be in such hands, for it is capable of being a great benefit to the public . . . We told the truth, for we were in the interest of the public; and the truth of that locality [Wall Street] was not complimentary in those days any more than it would be now." Hudson, *Journalism in America,* pp. 436–437.

78. Ibid., pp. 227, 384.
79. Parton, "New York Herald," p. 104. Bennett's statement is quoted in Pray, *Memoirs of Bennett,* p. 373.

2. "Time Past to Time Present"

1. The quotation in the title is from the *Canandaigua Repository and Advertiser,* 8 September 1845: "The telegraph . . . will doubtless become the common medium through which all important intelligence of a private or public nature must pass . . . The entire business community will jump at once from time past to time present in all their thoughts and transactions."
2. Jedidiah Morse to Samuel F. B. Morse, 8 February 1825, in Edward Lind Morse, *Samuel F. B. Morse: His Letters and Journals,* 2 vols. (Boston: Houghton Mifflin Co., 1914); see, e.g., Oliver W. Larkin, *Samuel F. B. Morse and American Democratic Art* (Boston: Little, Brown, and Company, 1954), pp. 77–79. Standard biographies of Morse are Samuel Irenaeus Prime, *The Life of Samuel F. B. Morse, Inventor of the Electro-Magnetic Recording Telegraph* (New York: D. Appleton, 1875); and Carleton Mabee, *The American Leonardo: A Life of Samuel F. B. Morse* (New York: Alfred A. Knopf, 1943).
3. These are the four basic elements of space in Stephen Kern's analysis, *Time and Space,* p. 3. The preoccupation with space in Jedidiah Morse's home is brought out vividly in Morse's watercolor painting "The Morse Family" (c. 1810) in which the family was portrayed scrutinizing a large globe. Reproductions of this painting appear in Larkin, *Morse,* facing p. 80, and in Brooke Hindle, *Emulation and Invention* (New York: New York University Press, 1981), p. 109.
4. "This sort of synthetic-spatial thinking is required in its most unalloyed form in painting or in sculpture where analytic, logical, verbal, or arithmetic thinking plays almost no role." Hindle, *Emulation and Invention,* p. 93.
5. Lee, *Daily Newspaper,* p. 492.
6. Everett M. Rogers, *The Diffusion of Innovations* (New York: The Free Press, 1962), pp. 81–87, presents a more elaborate structure of the characteristic stages of diffusion.
7. Here I followed the terminology of Brian Winston's study of media technology *Misunderstanding Media* (Cambridge, Mass.: Harvard University Press, 1986), pp. 16–27. The supervening necessity Morse per-

ceived and reacted to was apparently in the realm of social communications, and perhaps also the management of government affairs. (A perceptive discussion of this aspect appears in an unpublished paper by Richard R. John, Jr., "A Failure of Vision? The Jacksonians, the Post Office, and the Telegraph, 1844–1847.") Winston, however, proposes as the supervening necessity in the invention of the telegraph the demands of railroad technology. Although perhaps relevant to the British experience, such demands did not play an important role in the American context.

8. *Congressional Globe,* 27th Cong., 3d sess., 23 February 1843. See also Edward Lind Morse, "The District of Columbia's Part in the Early History of the Telegraph," *Records of the Columbia Historical Society* 3 (1900): 161–179; Poore, *Perley's Reminiscences,* vol. 1, pp. 309–310.
9. See Oscar Handlin's brief discussion of this aspect in "Man and Magic: First Encounters with the Machine," *American Scholar* 33 (Summer 1964), p. 415.
10. John W. Kirk, "Historic Moments: The First News Message by Telegraph," *Scribner's Magazine* (May 1892): 652–656.
11. Ibid.
12. Stephen Vail claimed in a letter to the *New York Sun,* quoted in *Scientific American,* 14 November 1891, supplement no. 828, that extras of Washington newspapers with the information received by telegraph were distributed at the depot. This is unlikely, as there was only a one-hour interval between the receipt of news by telegraph and the arrival of the train. Kirk, "Historic Moments," p. 655. (In *Electrical Review,* 7 January 1888, p. 6, the interval was considered an hour and a quarter.) A passenger on the train testified that the telegraphic news was printed on slips of paper and "scattered among the waiting crowds" (Kirk, p. 655). This may be the source of Vail's claim that extras were printed; yet the papers of both Morse and Vail (Library of Congress and Smithsonian Institution Archives respectively), which cover this period in detail, do not include any such slips of paper.
13. Kirk, "Historic Moments," pp. 653–655; J. Cummings Vail, *History of the First Telegraph Office in the United States,* pamphlet, New York Public Library. See also *Electrical Review,* 7 January 1888, 4 February 1888; *Scientific American,* 14 November 1891, supplement no. 828; *The Electrical Engineer,* 19 August 1891. Clippings can be found in the Donald McNicol Telegraph Collection, New York Public Library.
14. This date is preferred to the conventional 24 May 1844, according to the most reliable source, Alfred Vail's Journal, Vail Telegraph Collection, Smithsonian Institution Archives, and is supported by the *Baltimore Sun,* 25 May 1844. But see a telegraph slip in the Morse papers reading: "What hath God wrought," described in Morse's hand as the

first message over the line and dated by him 24 May, 8:45 A.M.; and J. Cummings Vail, *Early History of the Electro-Magnetic Telegraph* (New York: Hine Brothers, 1914), p. 17. Morse to Sidney Morse, 31 May 1844, Morse papers, describes the inauguration of the line but does not provide a date.

15. Alfred Vail's Telegraph Journal, Vail Telegraph Collection.
16. Alfred Vail's Telegraph Journal, 27–29 May 1844, Vail Telegraph Collection.
17. Ibid., 29 May 1844. Edward Lind Morse claimed that the Baltimore convention did not believe Wright's telegraphic dispatches, and they sent by rail a committee to confer with him and report back to them the following day. Morse, "District of Columbia and Telegraph," p. 174, and *Life and Letters*, vol. 2, p. 226. Vail's diary proves, however, that although a five-man committee was appointed, it negotiated with Wright by telegraph.
18. Samuel Morse, quoted in *Life and Letters*, vol. 2, p. 224. For Morse's emphasis on two-way correspondence in real time, see his report to the secretary of the treasury announcing the completion of the telegraph line to Baltimore, U.S. Congress, House, *Electro-Magnetic Telegraph*, H. Ex. Doc. 270, 28th Cong., 1st sess., 1844, p. 2. Compare Morse, "District of Columbia and Telegraph," pp. 175–176.
19. Virtually every account of the history of the telegraph describes this incident. *New York Herald*, 4 June 1844, is one of the earliest. And see *New York Tribune*, 2 November 1848; *New York Express*, 29 May 1844.
20. Clippings in Journalistic Series, vol. 1, O'Rielly Telegraph Collection, New York Historical Society.
21. Government Telegraph Account Books, Vail Telegraph Collection.
22. After the telegraph spread to connect important centers separated by great distances, its use expanded dramatically. Yet the patterns of its employment remained remarkably constant. A statistical account compiled by Marshall Lefferts demonstrated that in 1856 the press and the commercial community were the principal users of the lines connecting Boston and New York. Only about 6 percent of non-press messages were of a personal nature. Less than 3 percent of the non-press messages announced sickness or death. Messages for railroads constituted only 3 percent of all non-press messages. The statistics did not indicate that any federal or state authorities employed the line. (Such messages may have been included in the miscellaneous category which accounted for 17 percent of non-press messages.) Lefferts, "Electric Telegraph Influence," p. 259.
23. See, e.g., *Evening Express*, 29 June 1847; Reid, *Telegraph in America*, p. 308.
24. Since "the prices in one market immediately affect those of others . . .

without the very latest intelligence [brokers] could not act safely nor efficiently, either for themselves or their employers." The rationale for use of the telegraph by insurance companies was that they "must keep informed of the arrival, departure, and casualties of shipping at other seaports." Unidentified circular, entitled *The Electro-Magnetic Telegraph Considered as a Dividend-Paying Stock.* First Series, vol. 1, O'Rielly Telegraph Collection, New York Historical Society. See also *National Police Gazette,* 30 May 1846, for a variant but highly insightful analysis of the effects of the telegraph on commerce.

25. *The Electro-Magnetic Telegraph Considered as a Dividend-Paying Stock.*
26. Quoted in the *Philadelphia North American,* undated clipping, Journalistic Series, vol. 1, Henry O'Rielly Telegraph Collection.
27. Alonzo B. Cornell, *"True and Firm": Biography of Ezra Cornell, Founder of Cornell University* (New York: A. S. Barnes, 1884), pp. 94–95. "History of the Magnetic Telegraph," in *The Western Union Telegraph Company: Its Past, Present, and Future,* pamphlet (New York, 1883), p. 3. I could find no reference in the files of the *Herald* that indicated opposition to the telegraph.
28. My impression is that the *Herald* lagged behind the *Tribune* and especially the *Sun* in employing the telegraph. Broadly characterized, the *Herald* used the telegraph mostly for commercial news; the *Tribune* for political news; and the *Sun,* by far the greatest user, employed the telegraph extensively for reporting general news.
29. Elihu Katz, Jay G. Blumler, and Michael Gurevitch, "Utilization of Mass Communication by the Individual," in *The Uses of Mass Communications: Current Perspectives on Gratifications Research,* ed. Jay G. Blumler and Elihu Katz (Beverly Hills: Sage, 1974), pp. 19–31; Michael R. Real, "Media Theory: Contributions to an Understanding of American Mass Communications," *American Quarterly* 32 (1980): 238–258.
30. Helen MacGill Hughes, *News and the Human Interest Story* (Chicago: University of Chicago Press, 1940), pp. 2–19, 58; Hughes, "The Social Interpretation of News," *Annals of the American Academy of Political and Social Science* 219 (January 1942): 11–17; Hughes, *A Genealogy of Human Interest Stories* (reprinted from the *Journalism Quarterly,* March 1937). Compare the notion of the "narcotizing dysfunction" of mass media, Paul F. Lazarsfeld and Robert K. Merton, "Mass Communication, Popular Taste, and Organized Social Action," in *The Processes and Effects of Mass Communication,* ed. Wilbur Schramm and Donald F. Roberts (Urbana: University of Illinois Press, 1971), pp. 554–578.
31. James Gordon Bennett, quoted in Pray, *Memoirs of Bennett,* p. 373.
32. Clipping, Journalistic Series, vol. 1, Henry O'Rielly Telegraph Collection.
33. Elihu Katz and Daniel Dayan, "Media Events: On the Experience of

Not Being There," *Religion* 15 (1985): 305–314; Bernard Berelson, "What 'Missing the Newspaper' Means," in *Communication Research, 1948–1949,* ed. Paul F. Lazarsfeld and F. N. Stanton (New York: Duell, Sloan, and Pearce, 1949), pp. 37–47. See also David Paul Nord, "The Public Community: The Urbanization of Journalism in Chicago," *Journal of Urban History* 11 (August 1985): 411–441. The outstanding contemporary analysis of this function of the newspaper is in Parton, "New York Herald," p. 104.

34. James Gordon Bennett, quoted in Pray, *Memoirs of Bennett,* p. 373.
35. Morse to Francis O. J. Smith, 2 February 1839, Francis O. J. Smith papers, Maine Historical Society, Portland, Maine.
36. For short summaries of Kendall's career, see Arthur M. Schlesinger, *The Age of Jackson* (Boston: Little, Brown, and Company, 1945), pp. 67–70, and Glyndon G. Van Deusen, *The Jacksonian Era, 1828–1848* (New York: Harper Torchbooks, 1963), pp. 32–33.
37. See the Philadelphia *Spirit of the Times,* undated clipping; *New York Merchant Magazine,* quoted in *Regulations of the Atlantic Lake and Mississippi Telegraph Company,* undated pamphlet. Both appear in 1st series, vol. 1, O'Rielly Telegraph Documents.
38. Smith stands in the center of the study by Frederick Merk, with Lois Bannister Merk, *Fruits of Propaganda in the Tyler Administration* (Cambridge, Mass.: Harvard University Press, 1971). See, e.g., pp. 59–64. The quotation is from an address by C. J. Ingersoll in the House of Representatives, 27 April 1846 (p. 180).
39. Massachusetts, Suffolk, vol. 64, no. 1823 (20 December 1859), R. G. Dun and Co. Collection, Baker Library, Harvard University Graduate School of Business Administration, Cambridge, Mass.
40. Letter of dedication of the O'Rielly Telegraph Collection to the New York Historical Society. The printed letter appears as a frontispiece in each of the volumes of the collection.
41. Daniel J. Boorstin, *The Americans: The National Experience* (New York: Vintage Books, 1965), pp. 124–134.
42. The career of Henry O'Rielly is recorded in a monumental collection of letters, documents, and clippings he presented to the New York Historical Society. In some sixty volumes his journalistic activities, promotion of settlement in the West, and telegraphic enterprises are abundantly illustrated. The Henry O'Rielly papers at the New York Historical Society and the Rochester Historical Society include additional material which is especially interesting because it consists of that which O'Rielly did not see fit to include in the official record he assembled.
43. The patent rights for domestic employment of the telegraph were divided thus: Morse–9/16, Smith–4/16, the Vails–2/16, and Gale–1/16.

44. Based on an examination of telegraph acts of Connecticut (1846), Michigan (1847), Maine (1846, 1852), Maryland (1845), Massachusetts (1846, 1849, 1851), New Jersey (1853), and Pennsylvania (1847, 1848). All include a clause giving priority to state and federal authorities in the case of public emergencies, and the rights of eminent domain for telegraph lines.
45. O'Rielly to Meredith, February 1847, O'Rielly Telegraph Documents, vol. 4, New York Historical Society. Similarly, Kendall mentioned the "failure of all our efforts to raise a dollar in New York or Philadelphia." Amos Kendall to W. W. Corcoran, 7 January [?], Cyrus Field papers, New York Public Library.
46. Amos Kendall to W. W. Corcoran, 7 January [?], Cyrus Field papers, New York Public Library. The extensive use Corcoran and Riggs made of the Washington-Baltimore line is evident from the government telegraph account book, Vail Telegraph Collection.
47. It was O'Rielly who raised the investment funds from Swain. O'Rielly to Kendall, 9 January 1846; Kendall to O'Rielly, 11 January 1846; see also Swain to O'Rielly, 12 January 1846, O'Rielly Telegraph Documents, vol. 3.
48. O'Rielly to Kendall, 9 January 1846, O'Rielly Telegraph Documents, vol. 3.
49. It was common among editors who invested in the telegraph not to hold stock in their own name. Henry J. Rogers to Henry O'Rielly, 18 January 1846, 20 January 1846, and 19 March 1846; see also memorandum in O'Rielly's hand, 16 February 1846. O'Rielly Telegraph Documents, vol. 2. On Benjamin's move to Baltimore, see Hudson, *Journalism in America*, p. 589.
50. Henry O'Rielly to [?] Hart, 31 January 1846, O'Rielly Telegraph Documents, vol. 2.
51. Henry O'Rielly to [?] Meredith, February 1847, O'Rielly Telegraph Documents, vol. 4.
52. Ibid.
53. *New York Herald*, 8 July 1848.
54. Hudson, *Journalism in America*, p. 604.
55. *Scientific American*, 14 November 1891, supplement no. 828.
56. See, e.g., *American Telegraph System: Atlantic and Ohio Telegraph Co. Connecting the Eastern and Western States*, pamphlet (1845); *American Telegraph System, Great Central Range between the Atlantic and Mississippi, including the Ohio Valley and Lake Country*, pamphlet (1845); *American Telegraph System, Great Central Range between the Atlantic and Mississippi; Ohio River Telegraph—Pittsburgh, Cincinnati, Louisville, Connecting with the Atlantic and Ohio Line at Pittsburgh*, pamphlet (1846); *Atlantic Lake and Mississippi Telegraph, Pittsburgh, Cincinnati,*

and Louisville with Right of Extension to St. Louis, pamphlet (1846); Henry O'Rielly, *To the Citizens of St. Louis,* circular (1847); Henry O'Rielly, *To the Citizens of Cincinnati,* circular (1847); and numerous other pamphlets and circulars, 1st series, vol. 1, O'Rielly Telegraph Collection. For Kendall's criticism of O'Rielly's methods, see *Wheeling Daily Times,* 28 October 1847.

57. Quoted in Atlantic and Pacific Telegraph Range, Second Division West of the Mississippi, *Saint Louis and Fort Leavenworth Telegraph,* pamphlet (1847).
58. Quoted in Atlantic and Pacific Telegraph Range, *Telegraph to the Platte Country,* pamphlet (1847). The *St. Louis Daily Republican* treated that same sense of isolation with a measure of discreet intimidation when calling on residents of Platte country in Missouri to support the telegraph. It quoted a letter from that region, telling of a period in which "for several weeks we had no communication with the commercial cities, during which period a sudden rise took place in the price of Hemp, of which we remained in total ignorance. In the mean time, speculators arrived in the country, and taking advantage of our want of knowledge of the state of the markets, bought up all the Hemp in the country at a price not exceeding one half of what the article would have brought had we known the prices it bore in the cities east of us." The newspaper found that the loss these people suffered in one week, "from being destitute of such information as the telegraph could daily give them," was "more than enough to pay the whole cost of constructing a telegraph line."
59. William M. Swain to O'Rielly, 2 January 1846; O'Rielly to Meredith, February 1847; see also O'Rielly to Hart, 31 January 1846. O'Rielly Telegraph Documents, vol. 4.
60. Eliphelet Case to F. O. J. Smith, 12 November 1847 and 28 December 1847. F. O. J. Smith papers, New York Public Library.
61. *Philadelphia North American,* 21 May 1847.
62. Vail Telegraph Journal, beginning 27 May 1844, documents the sending of news, Vail Telegraph Collection; the reports are available in a comprehensive collection of newspaper clippings in Vail's scrapbooks, ibid.; and see *American Electro-Magnetic Telegraph: Its Early History as Shown by Extracts from the Records of Alfred Vail,* arranged by Cummings Vail (New York: Hines Brothers, 1914), p. 19.
63. Vail Telegraph Journal, 2 November 1844.
64. Thus John Wills, besides sending dispatches to newspapers, frequently sent messages to individuals. Telegraph Account Books, Vail Telegraph Collection. See also Wills's *Telegraphic Congressional Reporter* (Baltimore: John W. Woods, 1847), a telegraphic code that he probably printed to allow a large number of recipients to decipher his

news reports. This code was of the same nature, perhaps it was even the same code, that Alexander Jones used and described in *Historical Sketch of the Electric Telegraph including Its Rise and Progress in the United States* (New York: G. P. Putnam's Sons, 1852), pp. 123–132.

65. Unidentified clipping, 1st series, vol. 1, O'Rielly Telegraph Collection.

66. The terms "report," "reporting," and "reporter," used in the context of general newspaper news, became identified with telegraphic news gathering. Prior to the invention of the telegraph, the terms appeared to have specified verbatim accounts of official proceedings, such as of legislative bodies, courts, and speeches at public meetings, conventions, and so on. Since the news transmitted over the first telegraph line was characteristically of proceedings of official bodies (such as Congress, committees, and the Supreme Court), and these were traditionally referred to as "reports," that term appears to have been expanded with the spread of the telegraph from Washington, to encompass all news transmitted by telegraph, and ultimately, all descriptions of "fast," or breaking news (for example, of fires and wars). See *A Dictionary of American English* (Chicago: University of Chicago Press, 1944), p. 1931; *A Dictionary of Americanisms* (Chicago: University of Chicago Press, 1951); *Oxford English Dictionary,* pp. 474–476; Robert Lincoln O'Brien, "Machinery and English Style," *Atlantic Monthly* 94 (October 1904): 464–472; Marbut, *News from the Capital,* pp. 31–34, 57–61.

67. *Articles of Association of the New York and Boston Magnetic Telegraph Association, Together with the Office Regulations, Records of the Meetings of the Trustees, Stockholders, and Directors* (New York: Chatterton and Christ, 1848).

68. *Albany Evening Atlas,* 6 July 1846; Reid, *Telegraph in America,* p. 308. Early in January, when the first leg of the line between Albany and Utica was completed, the press along the route arranged for news reports in common. *New York Tribune,* 31 January 1846; for a detailed discussion, see Richard A. Schwarzlose, "The Nation's First Wire Service: Evidence Supporting a Footnote," *Journalism Quarterly* 57 (Winter 1980): 555–562.

69. See, e.g., *Rochester Daily American,* 20 November 1846; *Rochester Daily Democrat,* 5 January 1847.

70. Alexander Jones to O'Rielly, 18 January 1847; Rufus Chadwick to O'Rielly, 28 February 1847; Brooks to O'Rielly, 22 January 1847 and 26 February 1847. O'Rielly Telegraph Documents, vol. 5. See also *Pittsburgh Commercial Journal,* 23 January 1847; *Buffalo Commercial Advertiser,* 14 April 1847; *Rochester Daily Democrat,* 13 February 1847; *Albany Daily Knickerbocker,* 13 May 1847; and Schwarzlose, "Nation's First Wire Service."

71. *St. Louis Daily Union,* 30 November 1847; *Ohio State Journal,* undated clipping, Journalistic Series, vol. 1, O'Rielly Telegraph Collection; *Galena Gazette and Advertiser* and *Illinois Journal,* undated clippings, 1st series, vol. 1, O'Rielly Telegraph Collection.

3. "A Better-Organized System"

1. "The New York Associated Press, (embracing the Courier and Enquirer, Journal of Commerce, Express, Herald, Tribune and Sun,) [sees] the necessity, in the rapid extension and more general use of the telegraph lines throughout the United States, of introducing a more enlarged and better-organized system . . . for the early acquisition of important news, by telegraph, for publication." Daniel H. Craig, printed circular to managers of telegraph companies (New York, May 1851), MG 2, vol. 723, F2, no. 971, Public Archives of Nova Scotia.
2. The literature on the origins and dating of the New York Associated Press is elaborately reviewed in Schwarzlose, *Newsbrokers,* vol. 1, pp. 79–121.
3. Accounts which accept the 1848 dating include, among others: Simon N. D. North, *The Newspaper and Periodical Press: A Special Report for the Tenth Census* (Washington: Government Printing Office, 1884), pp. 106–107; James L. Crouthammel, *James Watson Webb: A Biography* (Middletown, Conn.: Wesleyan University Press, 1969), p. 91; Elmo Scott Watson, *A History of Newspaper Syndicates in the United States, 1865–1935* (Chicago, 1936), p. 1; Edwin Emery, *The Press and America: An Interpretive History of the Mass Media* (Englewood Cliffs, N.J.: Prentice-Hall, 1972), p. 199; Mott, *American Journalism,* p. 251; Victor Rosewater, *History of Cooperative News Gathering in the United States* (New York: D. Appleton and Co., 1930), pp. 64–66; Lee, *The Daily Newspaper,* p. 499; Oliver Gramling, *AP: The Story of News* (New York: Farrar and Rinehart, 1940), pp. 19–21; and Schwarzlose, *Newsbrokers,* vol. 1, pp. 79–121.
4. *New York Sun,* 29 November 1846. The *London Daily News* quoted therein estimated the *Sun*'s circulation at 51,000 copies. The *New York Herald,* 3 March 1848, reported the *Sun*'s circulation at 40,000 copies, more than double its own circulation of 17,280. Schwarzlose, *Newsbrokers,* vol. 1, p. 125. See also *New York Sun,* 2 June 1846.
5. *New York Sun,* 6 March 1848.
6. *New York World,* Sunday supplement, 20 January 1884.
7. In the interview, Beach stated that at the time "the telegraph had got as far South as Richmond, Va., and we employed a man named Bonsell to go to Richmond and receive the *Sun* dispatches at that point from the mail and from there send the news by wire." Beach's account

is inaccurate at this point. The telegraph line to Richmond was not completed until October of 1846.

8. The express was run by one person, Riddle (spelled "Riddell" in the *World* interview).
9. *New York Sun,* 29 May 1846; see also the issues of 8 and 11 June 1846.
10. The *Tribune* was probably not a party to the Mexican news agreement and did not participate in the meeting in the *Sun* office. In this newspaper interview Beach was referring to the parties to the subsequent agreements.
11. The *New York Sun,* 6 March 1848, stated that "the last great daily Pony Express, bringing all the news from Mexico and the South, from one to four days in advance of the mail, and the equally extravagant undertaking of bringing news from Halifax to this city, originated at the office of the New York Sun."
12. The text of the memo is as follows (indentation is unchanged):

    The original Telegraph News association in which was afterward merged the Harbor News Association was based upon a contract for a daily express to be run between Mobile and Montgomery Ala. by ___ Riddle during the existence of the Mexican War in 1846–7 the payment for which was conditioned upon a daily gain of 24 hours upon the regular mail.

    This contract was made by M S Beach in behalf of M Y Beach for the benefit of The Sun, but on its completion Mr Beach offered an equal interest to each of the other papers.

    This offer was first accepted and without a waver of hesitation by Mr Hallock for the Journal of Commerce.

    Mr H. J. Raymond then of the Courier accepted it next on behalf of that paper.

    Mr Hudson representing Mr Bennett asked time to present the matter to his principal after which he accepted it.

    Mr Brooks [of the *New York Evening Express*] followed Raymond with his acceptance.

    The memorandum is in the papers of Moses S. Beach, in the possession of Mr. Brewster Beach of New York City. My colleague Steve Jaffe brought this document, as well as several other important ones, to my attention and made it available for my use.
13. The pamphlet, by Henry O'Rielly, is quoted in full in the *American Telegraph Magazine* 1 (July 1853): 234–263. Beach's article is also cited (p. 261). The *American Telegraph Magazine* was an O'Rielly organ. Though edited by his son-in-law, Donald Mann, most of the journal was written by O'Rielly himself, as patently indicated by annotations in his hand on the copy in the O'Rielly Telegraph Collection, New York Historical Society.
14. The *New York Herald,* in later years, developed a journalism

historiography of its own that has greatly influenced historians of American journalism. The main postulate of the *Herald* tradition is that it led the New York press in news enterprise. In this context it also took credit for originating the NYAP. Bennett's statement to that effect in the *New York Herald,* 7 November 1866, set the tone for this tradition, which was reflected in William F. S. Shanks, "How We Get Our News," *Harper's New Monthly Magazine* (May 1967): 512–522, and later picked up and elaborated by Hudson, *Journalism in America,* pp. 366, 452, 609–610, and in Thomas B. Connary, "Great Business Operations: The Collection of News," *Cosmopolitan Magazine* 19 (1897): 21–32.

15. Beach's designation as the press's "generalissimo" was provided by F. O. J. Smith, in the context of the cooperative arrangements Beach initiated for news transmission from Boston. Francis O. J. Smith, "Vindication of the Telegraph," *New York Express,* 27 November 1846.
16. These include Charles A. Dana, who served at the time on Greeley's *Tribune,* quoted in James W. Wilson, *The Life of Charles A. Dana* (New York: Harper and Brothers, 1907), p. 485, and in Frank M. O'Brien, *The Story of the "Sun"* (New York: D. Appleton and Co., 1928), p. 118; the *New Orleans Daily Picayune,* 20 May 1866, clipping in the Beach papers, New York Historical Society; *Charleston Courier,* 11 May 1868, Beach papers. Alexander Jones, who served as the first "agent" (general manager) of the New York Associated Press from soon after its formation to 1851 also saw a connection between the Mexican War express and the emergence of the Associated Press. Jones, *Historical Sketch* p. 122. An analysis of Jones's fragmentary and contradictory account appears in my unpublished work "News Gathering in the Age of Systems"; and in "The News Frontier," pp. 110–112, 430–432. See Schwarzlose, *Newsbrokers,* pp. 95, 99, who rejects the possibility "that the AP originated during that conflict" on the authority of a scholar on Mexican War news coverage.
17. *New York True Sun,* quoting Philadelphia papers, 4 July 1846; and see 27 June 1846.
18. Amos Kendall to Thomas M. Clark, secretary of the Magnetic Telegraph Company, 25 June 1846, Amos Kendall papers, Library of Congress.
19. "The arrangement is shared by no less than six of our daily papers, and others doubtless might have been added to the list, had they been disposed to comply with the terms." *New York Mirror,* quoted in the *New York True Sun,* 29 June 1846; "We have it from the best authority that the opportunity was equally open to all." *New York Mirror,* quoted in the *New York True Sun,* 2 July 1846. Compare *Gazette and Times,* quoted in the *New York True Sun,* 2 July 1846.
20. *New York True Sun,* 27 June 1846; quotation from the *True Sun,* 2 July 1846.

21. *New York True Sun,* 27, 29, and 30 June; 1, 2, 4, and 7 July 1846.
22. *New York True Sun,* 27 June 1846; 2 July 1846.
23. *New York Express,* 2 July 1846; *New York True Sun,* 4 July 1846.
24. Even later, "New York Associated Press" was not a proper or official name, and it was used interchangeably with other designations, such as "the associated newspapers," "the New York press," "the associated press," and "the associated press of New York." (See, e.g., the usage in correspondence spanning 1848–1850 appended to Francis O. J. Smith, *An Exposition of the Differences Existing between Different Presses and Different Lines of Telegraph, Respecting the Transmission of Foreign News, Being a Letter and Accompanying Documents, Addressed to the Government Commissioners of the Nova Scotia Telegraph,* pamphlet, Boston, 1850, pp. 13–46.) As late as 1856 the association called itself the "General News Association of the City of New York."
25. The Magnetic Telegraph Company, *Proceedings,* pp. 65–68; compare *Regulations of the Magnetic Telegraph Company: Unalterable Rules,* p. 15. See also Jones, *Historical Sketch,* pp. 90–91.
26. "From the day [the telegraph line] opened and from then on, the language of the congressional accounts was the same in all papers. It was obvious that the same reporters were preparing them for all, and that probably a single telegraphic transmission supplied all of the subscribing papers in a given city." Marbut, *News from the Capital,* p. 79 and pp. 75–83. Richard A. Schwarzlose, in a study confined to three New York newspapers, reached a similar conclusion. He found that before the commencement of through telegraphic communications between Washington and New York, as early as 7 May 1846, an identical version of congressional reports by telegraph appeared in the *Herald* and the *Tribune.* These newspapers continued to receive common reports from Washington ever after, and in early July the *Courier and Enquirer,* the other newspaper he surveyed, began carrying these common reports as well. "Early Telegraphic News Dispatches: Forerunner of the Associated Press," *Journalism Quarterly* 51 (Winter 1979): 595–601.
27. *New York World,* 20 January 1884.
28. Schwarzlose, "Early Telegraphic News Dispatches."
29. Affidavit of Peter Cruikshanks, 12 November 1846, in the *New York Herald,* 20 November 1846.
30. Sworn statement of Alfred Ely Beach, 17 November 1846, quoted in the *New York Herald,* 20 November 1846. See also *New York Sun,* 18, 19, 21, and 25 November 1846. Beach's statement implied that the reports he prepared were sent seriatim to each of the newspapers he represented. As the published accounts in the various dailies were uniform, the procedure was intended either to extend the privilege of uninter-

rupted transmission (by sending for fifteen minutes to each newspaper), or to extend the interval during which the press held the news exclusively. See also *New York Journal of Commerce,* 18 November 1846; *New York Express,* 17, 20, 21, and 27 November 1846.

31. "Samuel Finley Breese Morse and the Electric Telegraph in Boston," *Bostonian Society* (December 1846): 31–55, clipping in the Donald McNicol Telegraph Collection, New York Public Library (microfilm). For John T. Smith's subsequent career, see Chapter 4 and report on John T. Smith, by R. G. Dun analyst, Massachusetts, Suffolk, vol. 66, no. 500 (22 May 1874), R. G. Dun and Co. Collection.
32. Craig's sworn statement is perhaps the single most illuminating description of the operations of the early New York Associated Press. It was published in the *Boston Evening Transcript,* 16 February 1859 (emphasis mine). That Smith was indeed responsible for telegraphing the steamer news to New York by November 1846 is established by the sworn deposition of Ira Berry, 19 December 1861, Smith papers, Maine.
33. F. O. J. Smith to Sidney Doane, president of the New York and Boston Telegraph Company (controlled by Smith), 22 February 1847, quoted in an anonymous, undated pamphlet, *Mr. Craig's "Superior Activity" in Telegraphing,* established by the Superior Court of New York as authored by Francis O. J. Smith, and published in 1851: See Daniel H. Craig, *The Convicted Libeller,* pamphlet (New York, 1863). The New York and Boston Magnetic Telegraph Association, from the beginning of its operations, charged fees on news dispatches according to the number of newspapers that published them. This was probably a reflection of a prevailing practice of newspapers sharing dispatches, rather than a brilliant anticipation of the formation of newspaper associations by two years. The New York and Boston Magnetic Telegraph Association, *Proceedings,* p. 14.
34. Variously spelled Francis Orman, Francis Osmond, even Francis Orville.
35. Both letters quoted in Francis O. J. Smith, *An Exposition of the Differences Existing between Different Presses and Different Lines of Telegraph, Respecting the Transmission of Foreign News, Being a Letter and Accompanying Documents, Addressed to the Government Commissioners of the Nova Scotia Telegraph,* pamphlet (Boston, 1850); Raymond's letter, pp. 27–28; Smith's reply, pp. 28–29. Craig's response, *A Review of an Exposition of the Differences Existing between Different Presses and Different Lines of Telegraph, Respecting the Transmission of Foreign News, Being a Letter and Accompanying Documents, Addressed to the Government Commissioners of the Nova Scotia Telegraph,* pamphlet (Halifax, 1950), pp. 18–19, includes

the full text of Raymond's letter, but only a fragmentary version of Smith's response, which omitted the key phrases quoted in the text.

36. There are numerous references in the press between 1846 and 1848 to the fact that the major New York dailies received news in common. See, e.g., *Journal of Commerce,* quoted in the *Baltimore American,* 25 November 1846; *New York Sun,* 4 August 1846, and the *Philadelphia Public Ledger,* several days later; the *Sun,* 18 November 1846, as well as other New York newspapers of the same date, and the *Buffalo Commercial Advertiser,* several days later; the *Philadelphia Public Ledger,* quoted in the *Philadelphia North American,* 9 February 1847, 21 May 1847; *Daily Commercial Journal,* 23 January 1847. All these sources can be found in clippings in Journalistic Series, vols. 1 and 2, Henry O'Rielly Telegraph Collection. Rosewater in *Cooperative News Gathering* cites several sources which indicate common telegraphic news arrangements prior to the spring of 1848, most notably: James Brooks to President Polk, 30 November 1847 (quoted p. 42); *Herald,* 27 November 1847 (p. 47); *Herald,* 15 January 1848 (p. 52). Rosewater and Schwarzlose's use of the *Herald*'s independent transmission of Henry Clay's speech of 13 November 1847 in Lexington, Ky., as proof that the NYAP did not exist at the time is based on the unwarranted assumption that the early association would have banned independent use of the telegraph by its members from any and all points. Schwarzlose, *Newsbrokers,* p. 95; but see his own statement, pp. 130–131.
37. *New York Herald,* 23 May 1846.
38. Unidentified clipping, Vail Telegraph Collection, and see similarly, the *New York Herald,* 24 February 1846.
39. Martin Van Creveld, "Technology and War," an early manuscript version of a book by that title (New York: The Free Press, 1989), made available to me by courtesy of the author. See also Stephen W. Littlejohn, *Theories of Communication* (Columbus, Ohio: Charles E. Merrill Publishing Co., 1978), pp. 30–75.
40. In an insightful discussion of the expected effects of telegraphy on news gathering, the *New York Herald,* 24 February 1846, appeared to grasp, to a certain extent, the implications of organization. It also anticipated that "private efforts must give way to [the telegraph's] great general power and efficiency, as a medium of conveying intelligence."
41. Amos Kendall, as postmaster general, summarized the implications of this structure in a report of 18 May 1838: "The number of newspapers published in the Union is believed to be from 1,300 to 1,500, and some publishers exchange with five or six hundred. Of these, not one hundred are of any practical use to them. They serve merely to overload and delay the mails, and are generally thrown away without being opened." Quoted in Kielbowicz, "News in the Mails," p. 340.
42. Timothy Pickering, quoted in Kielbowicz, "News in the Mails," p. 338.

43. "*Every* thing has to be read, examined, digested, and compared," complained Hezekiah Niles, editor of *Niles' Weekly Register*, discussing what he called the "editorial labor and drudgery" involved in exchange editing. Quoted in Kielbowicz, "News in the Mails," p. 347.
44. *Baltimore Sun*, 22 May 1846. See also *Philadelphia Dollar Newspaper*, 1 January 1846, and *Philadelphia North American*, December 1845, clipping in the O'Rielly Telegraph Collection, Journalistic Series, vol. 1; and in the Morse telegraph scrapbooks, Morse papers, Library of Congress (microfilm).
45. Quoted in Hudson, *Journalism in America*, pp. 533–534.
46. Parton, *Life of Greeley*, p. 137; see also *New York World*, 20 January 1884, for another coat story emphasizing the large pockets.
47. Schudson, *Discovering the News*, p. 16.
48. Quotation from Carlson, *Bennett*, pp. 124–125. The *Herald*'s circulation at the time is extrapolated from Hudson, *Journalism in America*, pp. 430 and 439.
49. Lee, *Daily Newspaper*, pp. 603–641.
50. Chamberlin, *Boston Transcript*, p. 82.
51. J. T. Scharf and Thompson Westcott, *History of Philadelphia, 1609–1884* (Philadelphia, 1884), p. 2002.
52. *The Evening Post: One Hundredth Anniversary*, pp. 45–47.
53. Parton, *Life of Greeley*, pp. 361–363.
54. Blondheim, "The News Frontier," pp. 420–423.
55. Based on Alfred Vail's detailed analysis of prices and capacity of telegraph lines in the summer of 1846. Alfred Vail to Postmaster General Cave Johnson, 26 November 1846, Vail Telegraph Collection.
56. Reid, *The Telegraph in America*, pp. 794–795. Compare Daniel H. Craig, *Answer of Daniel H. Craig to the Interrogatories of the United States Senate Committee on Education and Labor*, pamphlet (1883). The testimony can also be found in U.S. Congress, Senate, Committee on Education and Labor, *Report of the Committee of the Senate upon the Relations between Labor and Capital, and Testimony Taken by the Committee*, 48th Congress (1883–1885), vol. 2, pp. 1265–1283. The story is of course grossly exaggerated. The Old Testament in Hebrew contains 304,840 letters, which would require about 116 hours of transmission.
57. James D. Reid, "The Western Press and the Telegraph," *National Telegraph Review* 1 (October 1853): 234–238. See also Reid, *The Telegraph in America*, p. 794; Harlow, *Old Wires*, pp. 183–184; and Murat Halstead, "The Varieties of Journalism," *Cosmopolitan Magazine* 14 (December 1892), p. 204.
58. Receiving the last words of the president's message at Pittsburgh, the exhausted James Reid, after two sleepless nights, ended the transmission with his own personal touch: "God and Liberty! James K. Polk." Finding the Mexican battle cry conjoined to the president's name out-

raged some western editors, perplexed others, and inspired *Louisville Journal*'s George D. Prentice to one of his most memorable pronouncements: "What had James K. Polk to do with either God or liberty?"

59. Alfred Vail to Postmaster General Cave Johnson, 26 November 1846. Vail Telegraph Collection. This line had a policy that provided price reductions for the press. (All telegraph lines, however, had a policy of a sliding schedule of tariffs, making long messages cheaper per word than short ones.) Price reductions were warranted by the low occupation of that line prior to its connection to New York. An additional rationale for the special press rate was the policy of government support of the public press, and the quest of politicians responsible for policy for a good press.
60. Clipping, the *Rochester Democrat,* O'Rielly Telegraph Collection, 1st series, vol. 2.
61. Shanks, "How We Get Our News," p. 514.
62. The Magnetic Telegraph Company, *Minutes,* pp. 4, 12, 16; compare *Articles of Association of the New York and Boston Magnetic Telegraph Association, Together with the Office Regulations, Records of the Meetings of the Trustees, Stockholders, and Directors* (New York: Chatterton and Christ, 1848), p. 24; Atlantic Lake and Mississippi Telegraph Company, *Regulations,* broadside in the O'Rielly Telegraph Collection, New York Historical Society. This broadside, apparently dating to December 1845, was described by O'Rielly as "the first publication of Regulations for any telegraph line in America."
63. This analysis is based on Hudson, *Journalism in America,* p. 610.
64. Shanks, "How We Get Our News," pp. 514–515.
65. *New York World,* 20 January 1884.
66. This agreement is included in the Moses Beach papers. See also the *New York Herald,* 2 June 1848, 10 June 1848, and 25 January 1850. Clippings in the Morse telegraph scrapbooks, Morse papers.
67. Article 10 of the agreement provided for the sale of the association's news to out-of-town newspapers. This would of course imply a system of telegraphic arrangements, as the news, once published or prepared on slips, would not be salable. The lack of information on this subject probably implied the existence of a distinct telegraphic news arm operating on behalf of the association, or by news entrepreneurs connected with it.
68. By the time the Harbor News Association was formed, the five members had an additional naval arrangement in common. In May 1848, they chartered a steamer, the *Buena Vista,* to express European news from Halifax, the Cunarders' first port of call, to Boston, where the news could be conveniently telegraphed to New York (*Halifax Acadian Recorder,* 13 May 1848). There is no mention of this arrangement in the

harbor news agreement that was entered into after the *Buena Vista* began its Halifax-Boston schedule. This was probably because the steamer was chartered rather than owned outright. The procedures, like those of the early telegraphic news arrangements, were rather simple and direct and did not require much formal and legal detail.

69. Alexander Jones, who took part in a loose association of independent telegraph reporters, was probably working for the association from an early date. During this period, rather than merely supplying services to the association, he was apparently brought in as an employee, using his previous connections with news reporters on behalf of the New York Associated Press.
70. The *Tribune* joined the five associated newspapers in chartering the *Buena Vista,* though it was not a party to the Harbor News Association (*New York Herald,* 25 January 1850). It was in connection with putting the *Buena Vista* into service that Raymond, who together with Hudson formed the executive committee of the association, applied to F. O. J. Smith for a new deal for transmitting steamer news. Raymond found it necessary, on this occasion, to introduce the association and its membership, since it was altered by the addition of the *Tribune.* His main objective was to secure a period of uninterrupted transmission upon the arrival of the *Buena Vista* in Boston.
71. Harbor News Association, Agreement, 11 January 1849, Beach family papers; and also the Henry J. Raymond papers, New York Public Library; and the Manton Marble papers, Library of Congress. See also Richard A. Schwarzlose, "Harbor News Association: The Formal Origin of the AP," *Journalism Quarterly* 45 (Summer 1868): 253–260. By that time the *Buena Vista* had also been disposed of.
72. For Hudson's contradictory accounts and his 1848–1849 dating, see *Journalism in the United States,* pp. 366, 462, and 609–610.
73. Ibid. On Hudson's memory problems see Shanks, "How We Get Our News." Even an official document of the Western Associated Press took issue with Hudson's work shortly after its publication. Outraged by his history of the Associated Press, it claimed that "Mr. Hudson's book should be entitled: 'A History of the journalism of the New York *Herald;* with personal references to various no-bodies employed elsewhere on the Country Press.'" Western Associated Press, Office of the General Agent, printed circular (Chicago, May 1873), William Henry Smith papers, Indiana Historical Society, Indianapolis, Indiana.
74. Other evidence of the *Sun*'s leadership may be found in the service of the *Sun* as the association's perennial secretary (the *Journal of Commerce,* representing the Wall Street papers, was made president) and in the practice of measuring telegraphic reports by fractions of *Sun* columns.

75. Jones, *Electric Telegraph,* p. 138. The figure includes a large sum paid to the Canadian telegraph lines for securing exclusive transmission over their wires. Connery, "The Collection of News," estimated $21,000 as the total expense in 1848. Cogeshall, *Newspaper Record,* p. 146, estimated the telegraphic expenses of the New York Associated Press in 1851 to be $50,000, and U.S. Congress, Senate, S. Ex. Doc. 1, 32d Cong., 2d sess. (1853), p. 569, also established the cost of telegraphic dispatches to the Associated Press in 1852 at about $50,000. According to a detailed table of expenses prepared by Craig and published by the *New York Herald,* 16 July 1852, the telegraphic expenses of the association for the half year October 1851 to March 1852 amounted to $23,392. Marshall Lefferts, who was very close to the management of the Associated Press, estimated the yearly outlay in 1856 to be $60,000. Marshall Lefferts, "The Electric Telegraph: Its Influence and Geographical Distribution," *Bulletin of the American Geographical and Statistical Society* 2 (1857), p. 261. In 1851 Horace Greeley offered an estimate of $100,000 paid by the association in telegraphic tolls. Greeley, "Testimony," quoted in Hudson, *Journalism in America,* p. 541.
76. *Louisville Journal,* 6 November 1847.
77. Henry Luce supposedly rediscovered the value of slow news. David Halberstam, *The Powers That Be* (New York: Alfred A. Knopf, 1979), p. 61.
78. Quoted in Hudson, *Journalism in America,* p. 548.
79. Reid, "The Western Press and the Telegraph," p. 233.
80. *Philadelphia North American,* December 1845, clipping in Journalistic Series, vol. 1, O'Rielly Telegraph Collection.

4. "The Advantage of Time"

1. "I care little for the newspaper bravado of Mr. Craig . . . I had detected him in having cut the telegraph wires of the Boston and New York line, on the arrival of a foreign steamer, to give a *cypher* despatch addressed to Jacob Little of New York, the advantage of time." Statement of the defendant, 12 May 1859, *Daniel H. Craig v. Francis O. J. Smith,* Superior Court of New York, Office of the New York County Clerk, New York City.
2. The description of this incident is based on records of *Daniel H. Craig v. Francis O. J. Smith* in the Superior Court of New York, and of *Commonwealth of Massachusetts v. Francis O. J. Smith,* first in the Superior Court, Suffolk County, then in the Supreme Judicial Court of Massachusetts. The Judgment Roll and several depositions in the New York case are deposited in the Office of the New York County Clerk. Many other original depositions taken for the trial are in the Smith papers,

Maine Historical Society, Portland, Maine. The best sources for the Massachusetts case are interrogatories and depositions in the Smith papers, and several reports of the proceedings, published in pamphlet form, in the Smith papers, Maine, and in the Library of Harvard Law School, Cambridge, Mass. For a summary of the case in the Supreme Judicial Court of Massachusetts, see *Commonwealth v. Francis O. J. Smith,* 2 Allen 93 (Mass. 1865).

3. Smith had intimate relations with Annie and Carrie, the teenage daughters of John Kingsbury, a Maine innkeeper. Annie bore him a daughter, with whom she moved to New York, fearing Smith "was accessory in making way with" her. Deposition of John Kingsbury, 15 August 1864, Smith papers, Maine. Smith also had a long extra-marital affair with Mrs. Lucy P. De Maine, who bore him two children. Smith apparently arranged Lucy's marriage to one of his employees, Charles C. Northrup, when the woman who was to be his second wife learned of the affair. Thomas L. Gaffney, "Maine's Mr. Smith: A Study of the Career of Francis O. J. Smith, Politician and Entrepreneur" (Ph.D., diss. University of Maine, 1979), pp. 422–425, 484–491. Dr. Gaffney formerly of the Maine Historical Society, suggested that Smith's first wife, who was found dead shortly after learning of her husband's affair, committed suicide. In his will, Smith bequeathed a considerable sum to a home for destitute women.
4. This description is based on the market accounts of the *New York Herald* and the *New York Tribune* throughout November 1846. For a good summary see *New York Herald,* 27 November 1846.
5. *New York Herald,* 19 August 1846. "Speculator" was used at the time as a term both designating legitimate ventures—the profits of which were conjectural or risky—and indicating shady transactions.
6. *New York Herald,* 5 November 1846.
7. Ibid., 6 November 1846.
8. Ibid.
9. Ibid., 7 November 1846.
10. *New York Tribune,* 9 November 1846; *New York Herald,* 8 November 1846.
11. *New York Tribune,* 9 November 1846.
12. The *Herald* considered it the duty of the press to "endeavor to foil" the "herds of soulless speculators, who are taking advantage of the awful visitation of Providence that the people of Europe are at present suffering under, to make money and gratify their unholy cupidity." It called on citizens "north, south, east and west" to "be on their guard" and "beware of having any transactions with strangers, particularly those who appear anxious to operate in breadstuffs and provisions."

The newspaper concluded its public service announcement with a resounding "Take care!" *New York Herald,* 10 November 1846.

13. The woman inquired into the condition of the line and was told it wasn't working with certainty, and that the steamer news was expected any minute. She then asked about the condition of the New Haven line. Leaving the office briskly, she climbed into a chaise driven by the same man who had visited her in the telegraph office the previous November. The two were followed by the line inspector to the railroad depot and were seen boarding the four o'clock train to Worcester. After returning to the office, the inspector learned that the line had stopped working. With an assistant, he boarded the five o'clock train, detected and repaired a break in the line near Brighton, and reached the Worcester depot shortly after the slow four o'clock train had pulled in. In the Worcester telegraph office they found the woman making inquiries as to the condition of the line. She was informed that the line, repaired not long before, was employed in transmitting the foreign news for the New York Press. The woman and her accomplice left the office without sending a message, spent the night in Worcester, and returned to Boston the following morning. The man got off the train at Brighton. Smith, *Craig's Superior Activity; Craig v. Smith,* statement of the defendant, 12 May 1859, Office of the New York County Clerk; depositions of Ira Berry, 19 December 1861, 6 August 1864, and of Greenleaf Howe, 12 May 1860, 18 December 1861, 25 July 1864, Smith papers, Maine. *Journal of Commerce,* 24 and 25 February 1847.
14. The predominance of New Englanders among the pioneers of American news gathering and journalism is remarkable. Samuel Topliff and Harry Blake, both of Boston, were the acknowledged pioneers of systematic gathering of marine news. David Hale and Gerard Hallock, who intercepted ships for their news and also ran horse expresses to New York, were New Englanders, as were Horace Greeley, Henry J. Raymond, and the Beach family. Benjamin Day, the publisher of the first successful penny newspaper, the *New York Sun,* was from Connecticut. James Gordon Bennett, a Scotsman, began his newspaper career in Boston, and A. S. Willington, the acknowledged leader of news enterprise in the South, was also a Bostonian.
15. Secondary biographical accounts of Craig can be found in *Dictionary of American Biography,* vol. 2, p. 495; Reid, *Telegraph in America,* pp. 362–370, 794–795; Reid, "Western Press," pp. 242–243; *The Boston Herald and Its History* (Boston, 1878), pp. 5–6; Rosewater, *Cooperative News Gathering,* chap. 11; Thompson, *Wiring a Continent,* pp. 226–230; Harlow, *Old Wires,* pp. 184–185, 189–190; Carlson, *James Gordon Bennett,* pp. 199–200; J. T. Scharf, *History of Baltimore City and County* (Philadel-

phia, 1881), p. 8. The most comprehensive account is in Schwarzlose, *Newsbrokers*, vol. 1, pp. 169–170, 186–187.

16. The best source for Craig's Baltimore period is the records of the libel suit brought by Craig against Amos Kendall in the Supreme Court of New York. These records are deposited in the Office of the New York County Clerk. Also useful is Amos Kendall, *To the Stockholders of the American Telegraph Company*, printed circular (27 August 1860).
17. Craig, *Testimony*, p. 1266. For a significantly different account of the same event, see *Boston Herald and Its History*, p. 6. The most comprehensive account of this phase of Craig's career is "How the Associated Press News System Originated: Mr. Craig's Connection with the Association, etc." *New York World*, 11 December 1866. The account was probably written by Craig and served as the basis for his subsequent *Testimony*.
18. The letter, dated 12 March 1847, somehow found its way into the F. O. J. Smith papers, Maine.
19. Judgment Roll, *Craig v. Smith*, Superior Court, New York County, Office of the New York County Clerk. One important aftereffect of the case in the Superior Court of New York was a suit brought by the Commonwealth of Massachusetts against Smith for subornation while defending himself in the libel suit. Craig obtained a series of letters that passed between Smith and his "cast off mistress," Lucy De Maine, that included Smith's instructions as to how her husband was to testify in the trial. Smith was found guilty and sentenced to life imprisonment but succeeded in getting a motion for a re-trial, which never took place. Smith's main weapon in the argument for a re-trial was evidence of the very large compensation paid to Lucy De Maine for her willingness to expose Smith. (Part of which was use of a house Craig had built in Brooklyn, N.Y. See report on Craig, 3 April 1860, Westchester County, N.Y., vol. 1, p. 487, R. G. Dun and Co. Collection. See also entries for 28 June 1858, ibid., p. 250.) State of Massachusetts, Superior Court, Suffolk County, *Commonwealth vs. F. O. J. Smith, Opening Arguments of the Defendant* (February term 1866); *Commonwealth by Indictment vs. Francis O. J. Smith, Report of Testimony at the Trial in the Superior Court for Criminal Business for Suffolk County* (May 1865); State of Massachusetts, Superior Court, Suffolk County, *Commonwealth vs. F. O. J. Smith, Deft's Argument against the Power of the District Attorney to Enter a "Nolle Prosequi"* (January 1867); *Opening and Closing Arguments of Hon. Thomas Hayes and Hon. Francis O. J. Smith before the Superior Court of Suffolk County, Mass., on a Motion for a Retrial in the Case of the State vs. Smith*, pamphlet (Portland 1868); *Commonwealth of Massachusetts vs. Francis O. J. Smith on Indictment Procured by the Conspiracy and Perjuries of Daniel H. Craig and His Witnesses* (Portland, 1867). See

also Daniel H. Craig, *The Convicted Libeller,* pamphlet (New York, 1863).

20. Quoted in Atlantic Lake and Mississippi Telegraph, *Report of the Committee Appointed by the Citizens of Louisville to Examine the Subject of the Telegraph and the Responses of the Chairman of the Committee to the Hon. Amos Kendall's Attack upon the Committee,* pamphlet (1847), p. 16. See also *New York Herald,* 17 March 1847.
21. *New York Sun,* 18 November 1846. The affidavits are quoted in full in the *New York Herald,* 20 November 1846: in the *New York Express,* 17 November 1846; and in the *New York Journal of Commerce,* 18 November 1846.
22. *New York Sun,* 25 November 1846.
23. Ibid., 21 November 1846. For Smith's response see the *New York Journal of Commerce* and the *New York Express,* 20 November 1846. Smith's comprehensive, documented statement appeared in the *New York Express,* 27 November 1846.
24. *New York Herald,* 22 November 1846.
25. Ibid.
26. Ibid., 20 November 1846.
27. Clipping in Miscellaneous Series, vol. 3, Henry O'Rielly Telegraph Collection.
28. Thompson, *Wiring a Continent,* pp. 327–328; see also Chapter 5.
29. Based on Ross M. Robertson, *History of the American Economy* (New York: Harcourt, Brace, and World, 1964), pp. 348–358. Quotation from p. 348.
30. Craig later claimed that until the telegraph extended to St. John, "I had the field to myself." Craig, *Testimony,* p. 1266. James D. Reid also considered Craig to have been "in undisputed possession of . . . a very valuable line of trade." *Telegraph in America,* p. 363. Reid, however, also states that Smith assumed "the right to refuse Craig's news reports until the regular press dispatches were transmitted," and Craig confirms this fact. Craig, *Review,* pp. 18–19. Although there clearly were obstacles and restrictions on Craig's use of the wires, he could still, obviously, supply the Boston press in advance of the NYAP, and he could also devise ways of telegraphing his news reports to New York by proxy.
31. Reid, *Telegraph in America,* p. 363; and Craig, *Review,* p. 18.
32. Craig was virtually forced to supply the NYAP when he had the foreign news exclusively, "*hours* in advance of all others, and hours in advance, even, of the arrival of the English steamer." Because of the "*cunning knavery* of Smith and a subservient slave or *understrapper* of his of the name *Saddler* [superintendent of the line] . . . I could never get a word of news off to New York unless it was addressed to the

Associated Press." Craig, *Review*, pp. 18–19. Craig, however, coveted the post of foreign news agent of the NYAP. Craig, "To the Public," *Boston Daily Mail*, 2 January 1850.

33. Reid, *Telegraph in America*, p. 364.
34. Smith to Raymond, 15 May 1848, appended to Smith, *Exposition*, pp. 28–29.
35. The *Buena Vista* arrived in Halifax on 8 May after a 6-day journey from New York. "A Great Project and Failure," and "Shipping—Port of Halifax," *Halifax Acadian Recorder*, 13 May 1848; and 27 May and 10 June 1848. See also *New York Herald*, 25 January 1850; Arthur L. Johnson, "The Boston-Halifax Steamship Lines," *The American Neptune* 37 (1977), p. 233. Secondary accounts of the *Buena Vista* episode include Thompson, *Wiring a Continent*, pp. 226–227; Rosewater, *Cooperative News Gathering*, pp. 66–68; Schwarzlose, "Early Telegraphic News Dispatches"; and more recently and comprehensively, *Newsbrokers*, vol. 1, pp. 96–99, 102–104.
36. When it arrived at Halifax on 6 June 1848, the *Buena Vista* carried only five passengers. "Shipping—Port of Halifax," *Halifax Acadian Recorder*, 10 June 1848.
37. Craig, *Testimony*, p. 1266; Reid, *Telegraph in America*, p. 363; Thompson, *Wiring a Continent*, p. 227.
38. Minutes of the Montreal Board of Trade, 5, 8, 11 September 1846; 6, 7, 21 October 1846; 26 December 1846; Report of the Council, 31 March 1847, Records of the Montreal Board of Trade, MG 28 III 44, National Archives of Canada, Ottawa.
39. Frederick A. Wilson, Secretary, Montreal Board of Trade, to D. Daly, Provincial Secretary, 4 November 1846, New Brunswick, House of Assembly, *Journal* (1847), pp. 31–32; Nova Scotia, House of Assembly, *Journal* (1847), pp. 618, 641, 642, 697, 698, 709, 715, 717, and appendixes 8, 33, 85; *Journal* (1848), pp. 69, 93, 108, and appendixes 22, 36. See also British North American Electric Telegraph Association to J. S. Saunders, Provincial Secretary, 15 March 1847, New Brunswick Executive Council, MG 9, A 1, vol. 53, p. 1547, Public Archives of New Brunswick, Fredericton, NB (microfilm).
40. John Harvey to Earl Gray, 11 March 1848, in Nova Scotia, House of Assembly, *Journal* (1849), appendix 28, pp. 209–210; and George R. Young and Michael Tobin, "Memorandum Relative to the Electric Telegraph," 19 June 1848, *Journal*, pp. 211–214.
41. Lawrence R. Darrow to Michael Tobin and George R. Young, 12 June 1848, *Journal* (1848), p. 219; Daniel H. Craig, "To the Agent of the Ass. Press," 10 May 1859, circular in Nova Scotia Telegraph Company, *To the Public*, pamphlet (1859).

42. Deposition of Greenleaf Howe, 18 December 1861, Smith papers, Maine.
43. Daniel H. Craig, "Sworn Statement to the Committee on Mercantile Affairs and Insurance," 14 February 1859, quoted in the *Boston Transcript,* 16 February 1859.
44. Craig, "Sworn Statement."
45. Lawrence R. Darrow to J. S. Saunders, Provincial Secretary, 8 January 1849, New Brunswick Executive Council, MG 9, A 1, vol. 53, pp. 1553–1555, Public Archives of New Brunswick.
46. The account of the NYAP's role in establishing the New Brunswick telegraph in Reid, *Telegraph in America,* pp. 341–342, is inaccurate.
47. Craig, *Testimony,* p. 1266.
48. Craig, "Sworn Statement"; *Halifax Nova Scotian,* 5 March 1849. The experience apparently left a deep impression on Craig. He once remarked that the effort involved in reading the whole of a thick pamphlet written by Smith was a "toil compared with which ice-cutting in the Bay of Fundy may be called an exhilarating occupation." *The Convicted Libeller,* p. 3.
49. Craig, *Testimony,* pp. 1266–1267. The AP obviously had to pay for the transmission of John T. Smith's message as well. Its contract with the eastern telegraph companies stipulated that they had to employ a "good and sufficient relay of express from Halifax to St. John's [*sic*]." They were liable to pay if the express failed to arrive as a result of "carelessness . . . want of energy, or unnecessary delay, or by reasons of an imperfectly arranged and organized express." Agreement between L. R. Darrow and H. J. Raymond and Frederic Hudson, 9 February 1849, appended to Smith, *Exposition,* pp. 30–31. The express probably cost Craig more than $1,000. *New York Herald,* 25 January 1850; *New York Courier and Enquirer,* May 1849, quoted in John W. Regan, "The Inception of the Associated Press," *Transactions of the Nova Scotia Historical Society* 19 (1912), p. 6.
50. *Halifax British Colonist,* 8 March 1849, quoted in Regan, "Inception," p. 9. Compare "Rivalry" and "The Express," *Halifax Nova Scotian,* 12 March 1849, and 19 March 1849; *Halifax Acadian Recorder,* 10 March 1849, and 17 March 1849; "How the AP Originated," *New York World,* 11 December 1866.
51. L. R. Darrow to F. O. J. Smith, January 1850, appended to Smith, *Exposition,* pp. 22–23.
52. Craig, "Sworn Statement"; "To the Public"; *Testimony,* p. 1267; Craig, *To the Members of the New York Associated Press,* printed circular (November 1866), Manton Marble papers, Library of Congress; "How the AP Originated," *New York World,* December 1866.
53. This conflict is best documented in the voluminous collection of clip-

pings, documents, and letters in the O'Rielly Telegraph Collection. Especially useful are Journalistic Series, vols. 2 and 3; 1st series, vol. 2; and Miscellaneous Series, vols. 1–3. For a summary see Thompson, *Wiring a Continent*, pp. 97–193.

54. O'Rielly elucidated the connection between his telegraphic activities and his political credo in a detailed statement concerning his relations with the NYAP published in the *American Telegraph Magazine* 1 (July 1853): 234–261.
55. See, e.g., Atlantic Lake and Mississippi Telegraph Range, *New Telegraphs, Reduced Rates, etc.* (1848); Atlantic Lake and Mississippi Telegraph, *A Brief Sketch of the Origin and Progress of the Telegraph Lines under the Management of Henry O'Rielly* (October 1848); Atlantic Lake and Mississippi Telegraph Range, *New York and New England Telegraph;* Atlantic Lake and Mississippi Telegraph Range, *Equal Rights and No Monopoly in Telegraphy; The American Newspaper Press versus the Threatened Telegraph Monopoly of F. O. J. Smith and Kendall* [sic] *Controllers of Morse's Plan* (1848). All of the above can be found in the O'Rielly Telegraph Collection.
56. "They pay the newspapers liberally for puffing them, and nearly every newspaper in all the West is in their interest." John J. Speed to Smith, Smith papers, Maine. Wisely, perhaps, O'Rielly and his newspapers spared Morse the worst of the abuse. He was depicted either as being above the fray or, more characteristically, as being misled by the evil counsel of his chieftains.
57. Clipping in Journalistic Series, vol. 2, O'Rielly Telegraph Collection, and quoted in part in the *Pittsburgh Mercantile Advertiser,* 8 July 1848.
58. *New York Observer,* 7 October 1848; for comment, see *New York Tribune,* 10 October 1848.
59. See, e.g., *New York Star,* 16 May 1848.
60. "An Act to Provide for the Incorporation and Regulation of Telegraph Companies," 12 April 1848, New York, *Laws* (1848). I have examined telegraph acts of Maryland (1845), Maine (1846, 1852), Connecticut (1846), Michigan (1847), Pennsylvania (1847, 1848), Massachusetts (1849, 1851), and New Jersey (1853). Although most include a clause giving priority to state and federal authorities in the case of public emergencies, New York was alone in giving priority to the public press. See *New York Tribune,* 18 March 1850.
61. O'Rielly at first experimented with employing the Zook and Barnes Columbian Telegraph over segments of his Kentucky lines, but it was shortly ruled an infringement on Morse's patent.
62. Smith was not overly concerned with Downing's prospective line, and he demonstrated this unconcern by what was probably the best publicity stunt of his telegraphic career. In the *Boston Evening Transcript,*

he challenged the House interests to wage $1,000 against his own on a race between their telegraph machine and his 2,500-pound Durham Bull in carrying 1,000 words from New York to Boston. The item was an instant hit with exchange editors throughout the country. Although some argue that the ploy ultimately backfired by giving much publicity to the House telegraph and encouraging subscriptions, the challenge did have its merits. Smith's original challenge was issued on 27 October and published in the *Boston Evening Transcript* on 28 October 1847. Numerous references to the challenge appear in O'Rielly's Journalistic Series, and in Morse's scrapbook, Morse papers, Library of Congress (microfilm). When the line was finally completed in the fall of 1849 it proved so unreliable that it posed no serious competition to Smith's line. Reid, *Telegraph in America*, pp. 360–361; Thompson, *Wiring a Continent*, pp. 170–172.

63. *Scientific American*, 14 October 1848; *New York Herald*, 27 December 1848; *New York Journal of Commerce*, quoted in the *Louisville Morning Courier*, 3 May 1848. But see also Cornell to *Journal of Commerce*, 30 November 1848; and "Morse's Telegraph and Baine's," 9 November 1848, unidentified clippings, Morse scrapbooks.
64. A copy of the agreement between O'Rielly and the Bain patentees can be found in 1st series, vol. 1, O'Rielly Telegraph Collection; "Certificate of the Formation of the New York and New England Telegraph Company," 23 April 1849, Smith papers, Maine. Quotation from Reid, *Telegraph in America*, p. 361.
65. Henry O'Rielly to the Associated Press of the City of New York, 23 January 1850. Quoted in O'Rielly, "Associated Press," p. 241.
66. Smith to Kendall, 26 March 1850 and 10 April 1850; Kendall to Smith, 22 April 1850; Smith to B. B. French, 1 May 1850, Smith papers, Maine; Smith, *Exposition*, pp. 3–4; Craig, *Review*, p. 17; see also *Boston Daily Mail*, 26 January 1850.
67. Craig to John T. Smith, 14 December 1849, appended to Smith, *Exposition*, pp. 36–38; Craig, "Statement"; Craig, *Review*, p. 14; Reid, *Telegraph in America*, p. 364.
68. Quoted in Craig, *Review*, p. 14; see also Gerard Hallock et al., "The European News—To the Public," *New York Herald*, 25 January 1850; and Craig, "Statement."
69. Craig, "Statement."
70. Proceedings of the Nova Scotia House of Assembly, 6 February 1850, in the *New York Herald*, 1 March 1850; see also *Halifax Sun*, 16 February 1850, quoted in the *Herald*.
71. The description of the contest between Craig and John T. Smith is based largely on Reid, *Telegraph in America*, pp. 364–366. Reid had the

"private correspondence of Craig with Hudson and Raymond" as his source (p. 366).

72. Craig to Frederic N. Gisborne, Superintendent of Nova Scotia Telegraph, RG 1, vol. 442, no. 81, Public Archives of Nova Scotia, Halifax, NS; Hallock et al., "European News"; Craig, *Review,* p. 4; Craig, "Statement."
73. E. B. Foster to the Proprietors of the Magnetic Telegraph, 27 November 1849, appended to Smith, *Exposition,* p. 34. For Smith's involvement see Jones, *Historical Sketch,* p. 140; Craig, "To the Public"; Craig, *Review,* p. 5; and Charles O. Rogers to Smith, 6 September 1852 and 11 September 1852, Smith papers, Maine.
74. *Boston Evening Transcript,* 7 December 1849; Craig, *Review,* p. 4.
75. Reid, *Telegraph in America,* pp. 362, 794.
76. Ibid., p. 366.
77. Smith's company was chartered under the general telegraph law of the state of New York; and the AP succeeded in having the act include a clause that justified priority to the public press in its use of the wires. Smith construed the clause strictly to mean that priority should be granted only if the entire press were united. With the split in the Boston association, that condition was no longer fulfilled.
78. Darrow to Smith, 20 December 1849, appended to Smith, *Exposition,* pp. 20–21.
79. Deposition of James Stockford, 14 October 1861, Smith papers, Maine; *Halifax Acadian Recorder,* 12 January 1850; *St. John Courier,* 22 December 1849, in the *New York Herald,* 30 December 1849; *Boston Evening Transcript,* 29 and 31 December 1849, and 2 January 1850. The *New York Herald,* 31 December 1849, includes a full statement by Darrow of previous interferences in the transmission of steamer news over the New Brunswick line.
80. See, e.g., *Boston Evening Transcript,* 31 December 1849. The quotation is from Craig, *Review,* p. 16.
81. Ezra Cornell to J. J. Speed, 28 April 1847, Smith papers, New York Public Library. (Norton was the same operator to whom Smith's line gave priority over the AP in November 1846. See Norton's response in the *New York Express,* 21 November 1846.) See also Craig, "To the Public"; *Buffalo Evening Express,* 29 June 1847; *Belfast Journal* (Maine), quoted in the *Albany Journal,* and then in an unidentified newspaper clipping, July 1847, Journalistic Series, vol. 2, Henry O'Rielly Telegraph Collection. A somewhat unconventional source is an essay composed by Craig as a practice exercise for using the "Rapid" telegraph system he promoted. In it he described the operation of the visual telegraph between Halifax and Boston. *Craig's Manual of the Tele-*

graph Illustrating the Electro-Mechanical System of the American "Rapid" Telegraph Co. of New York (New York: John Polhemus, 1879), p. 94.

82. Craig, "To the Public."
83. Nova Scotia, House of Assembly, *Journal* (1850), pp. 466, 506, 523, 580, and appendix 83, pp. 246–252.
84. Thompson, *Wiring a Continent,* p. 174.
85. Smith to Hudson and Raymond, 31 December 1849, quoted in Smith, *Exposition,* p. 13; Smith to the Government Commissioners of the Nova Scotia Telegraph Line, 29 December 1849, unidentified clipping quoting also Gisborne's response to Smith's letter (31 December 1849), Morse scrapbooks, Morse papers (microfilm).
86. Craig, "Statement."
87. After Craig was beaten by Smith to the Nova Scotia telegraph, he apparently had his message sent over the Bain line, already in operation. Lefferts to O'Rielly, 11 December 1849, O'Rielly Telegraph Documents, vol. 5. In "Associated Press," p. 240, O'Rielly stated that the line was completed in September 1849, and he quoted an article from the *New York Journal of Commerce,* 29 September 1849, to that effect. Yet the AP editors, as late as 24 January 1850, stated that "between us and the news there are 350 miles of telegraph wire, over which the press are not permitted to send a message." Hallock et al., "European News." The *New York Herald,* 15 January 1850, referred to experiments in sending press messages over the Bain line, and as late as 17 February 1850 the newspaper expressed "gratification and to a certain degree astonishment" at the satisfactory operation of the Bain line. It may have been that the line, completed in September 1849, was not working reliably until early in 1850.
88. O'Rielly to the Associated Press of the City of New York, 23 January 1850, quoted in O'Rielly, "Associated Press," p. 241.
89. The express was arranged by Winslow and Co. On one occasion it covered the distance in as little as three hours and twenty minutes. Unidentified clipping, 9 February 1850, Morse scrapbooks.
90. Gisborne to the Commissioners of the Electric Telegraph, 26 January 1850, RG 1, vol. 442, no. 84; Gisborne to Joseph Howe, 9 March 1850, RG 1, vol. 442, no. 90, Public Archives of Nova Scotia; *Halifax Nova Scotian,* 11 and 18 February 1850.
91. Kendall to James Gordon Bennett, 12 April 1850, quoted in the *New York Herald,* 13 April 1850. See also [Nova Scotia Telegraph] Commissioners to Darrow, 18 March 1850; Darrow to Commissioners, 19 March 1850, RG 31, series 120, vol. 10, Public Archives of NS.
92. There were also attempts to sabotage the construction of O'Rielly's competing line between Boston and Portland. *Boston Daily Mail,* 22 March 1850, quoted in an unidentified clipping, Morse scrapbooks.

93. Subscription Book to the Stock of the Maine Telegraph Company, Smith papers, Maine. Subsequently, on 29 July 1851 Alfred Vail sold his stock in the company to Hiram O. Alden, a large stockholder and president of the company; yet Morse apparently held on to his own stock. Vail to Alden, 29 July 1851, Alden papers, Maine Historical Society.
94. Kendall to Smith, 2 April 1850, Smith papers, Maine; also quoted in the *New York Herald,* 13 April 1850; see also Smith's comments in reply in the *New York Daily Globe,* 18 April 1850.
95. Kendall to Smith, 1 April 1850, Smith papers, Maine. The *New York Herald,* 24 January 1850, similarly stated that "we suspect . . . that the legal contest, now before the courts, in which Prof. Morse has a deep interest, will be prejudiced very materially in the minds of the community, and in the minds of the highest judicial authorities, in consequence of the impudent, overbearing, tyrannical, and abusive conduct of Fog Smith against the New York Press."
96. *New York Herald,* March 1850, clipping in the Morse scrapbooks.
97. Exclusive control of the wires gave Craig a small field for private enterprise as well. He established, in conjunction with the Nova Scotia telegraph, a system for forwarding private business messages between American businessmen and their British counterparts, employing the organizational machinery of the AP and utilizing its priority over the wires. Huntley Gisborne to Craig, 17 May 1852; Marshall Lefferts et al., "Public Notice," June 1852; Craig to Halifax Telegraph Company, 8 June 1852, Nova Scotia Telegraph Correspondence, 1850–1852, RG 31, series 120, vol. 10, no. 10, Public Archives of Nova Scotia; see also *Nova Scotian,* 26 November 1849.

5. "A Monstrous Monopoly Is Created"

1. "The agent of the Associated Press in New York really *edits* the news of the world for this country; can give the public what he pleases and withhold what he pleases, and distribute it to whom he pleases—and thus a monstrous monopoly is created—of which it is sufficient to say, that it is liable to very great abuse, and that it implies too much power to be entrusted to any one man, or to any one company." Massachusetts, Senate, Committee on Mercantile Affairs and Insurance, testimony of George Lunt, 9 February 1859, quoted in the *Boston Courier,* 10 February 1859.
2. Daniel H. Craig, *To Editors,* printed circular (New York, 9 November 1866), Manton Marble papers, Library of Congress.
3. Daniel H. Craig, *To the Members of the New York Associated Press,* printed circular (New York, November 1866), Manton Marble papers.

4. This goal is what Richard Schwarzlose calls "newsbroking." According to this definition, a newsbroker or newsbrokerage is involved in "the daily collection and distribution of news dispatches via communication systems among journalists in several communities." Schwarzlose, *Newsbrokers,* vol. 1, p. ix. Prior to the application of Craig's plan of 1851, the NYAP could hardly be considered a newsbrokerage in the sense of a daily distributor of news "in several communities."
5. Robert W. Russell, *Remarks of R. W. Russell, One of the Executive Committee of the American Telegraph Company, in Reply to the Statements of Messrs. Abram S. Hewitt, Cyrus W. Field, Henry J. Raymond, and Others, Made at the Meeting of Stockholders, on June 29th, 1860,* pamphlet (New York: William C. Bryant and Co., 1860), p. 56.
6. Carr and Johnson to President and Directors of the Provincial Telegraph Lines, 20 November 1850, RG 1, vol. 442, no. 88, Public Archives of Nova Scotia.
7. Nova Scotia Telegraph Commissioners to Carr and Johnson, 27 December 1850, Nova Scotia Telegraph Correspondence, 1850–1852, RG 31, series 120, vol. 19, no. 10, Public Archives of Nova Scotia. See O'Rielly, "Associated Press," pp. 252–260.
8. There were three prominent news agencies at the time, those of Abbot and Winans, Carr and Johnson, and John Wills. These agencies apparently cooperated and exchanged news, while serving different regions. It was probably to the combination of these agencies that Craig referred when stating that "more than three quarters of all the Southern, Eastern, and Western journals were firmly banded together in an opposition organization." Craig, *To the Members.* See O'Rielly, "Associated Press," pp. 248–257, for the subsequent concerted movements of these organizations to oppose Craig's control of the foreign news; and compare Craig, *Letter to Freeman M. Edson,* pamphlet (New York, 1853), pp. 15–19. For an elaborate discussion of Abbot and Winans's operation, see Schwarzlose, *Newsbrokers,* vol. 1, pp. 85–87, 118–119, 171–178.
9. The *Boston Evening Transcript,* 11 February 1850, prefaced its foreign news report: "European News . . . from Halifax, communicated to the morning papers, 'the only ones who now receive it, and pay for it, and the only ones entitled to any credit for the same.'" In its previous issue the *Transcript* introduced the foreign news dispatch with Craig's statement that it was "offered for transmission over the wires with the understanding on his part that the different lines will protect the Associated Press in their just and legitimate rights." Ibid., 9 February 1850.
10. Craig, *Manual,* p. 94; Craig, *Testimony,* p. 1268. The number of vessels docking first at New York was increasing rapidly throughout the

1850's relative to the number docking first at Halifax. After the establishment of the Collins service, and the inauguration of direct journeys to New York by Cunard in 1848, the importance of intercepting steamers at the port of New York increased significantly. Albion, *New York Port*, pp. 303–335. By 1859 there were fourteen scheduled steamer arrivals at the port of New York, compared with only two at Boston with a stop at Halifax. *Debate in the Massachusetts Senate, of the Bill to Incorporate the Boston and Halifax Telegraph Cable Co.*, 24 and 26 March 1859 (Boston: Henry W. Dutton and Son, 1859), p. 28.

11. Craig, printed circular to managers of telegraph companies (May 1851). A copy addressed to J. R. Young (one of the Nova Scotia telegraph commissioners) is in MG 2, vol. 723, F2, no. 971, Public Archives of Nova Scotia; another copy is in the Manton Marble papers. The circular is quoted in Reid, "Western Press," pp. 243–246.
12. Ibid.
13. Ibid.
14. The only subsequent American wire service that operated on an entirely different plan was the Press Association of the Confederate States of America. The major European wire services had a very different structure. Francis Williams, *Transmitting World News* (Paris: UNESCO, 1953); Graham Storey, *Reuter's Century, 1851–1951* (London: Max Parrish and Co.); Clifford F. Weigle, "The Rise and Fall of the Havas News Agency," *Journalism Quarterly* 19 (September 1942): 277–286; Fritz Stern, *Gold and Iron: Bismarck, Bleichroder, and the Building of the German Empire* (New York: Vintage Books, 1979), pp. 262–279; Robert W. Desmond, *The Information Process: World News Reporting to the Twentieth Century* (Iowa City: University of Iowa Press, 1977).
15. For the development of the function of reporters, see, e.g., Thomas Leonard, *The Power of the Press: The Birth of American Political Reporting* (New York: Oxford University Press, 1986), pp. 77–96. James W. Carey points out the significance of this development in "The Problem of Journalism History," *Journalism History* 1:1 (1974): 3–5, 27–28; and "'Putting the World at Peril': A Conversation with James W. Carey," *Journalism History* 12:2 (Summer 1985): 38–53.
16. The best descriptions of the system are in Craig's circular to managers of telegraph companies and his various pamphlets and circulars. See, e.g., "The Associated Press," *Boston Evening Gazette*, 13 November 1858. For later descriptions of the news system that Craig organized, see Shanks, "How We Get Our News"; William Alpin, "At the Associated Press Office," *Putnam's Magazine* (July 1870): 23–30; "The Metropolitan Newspaper," *Harper's New Monthly Magazine* 56 (December 1877): 43–59. For a summary, see Rosewater, *Cooperative News Gathering*, pp. 89–91.

17. The origins of Craig's telegraphic news scheme are best understood against the background of developments in the management of the major New York dailies and the earliest experiments in systematizing the fledgling telegraph industry. Craig, operating on the borderline between the press and the telegraph, incorporated elements of both institutions into the foundations of his system. The challenge of transmitting the foreign news through to New York, requiring the coordination of a sequence of separate lines and then the subsequent distribution of the news across the country, stimulated Craig and his close ally Marshall Lefferts to aim at integrating the various lines, at least functionally, into a system. Craig to Lefferts, 11 October 1850; Lefferts to Craig, 26 October 1850; Craig to Lefferts, 5 December 1850, quoted in Marshall Lefferts, *Address to the Stockholders of the New York and New England Telegraph Company,* pamphlet (New York: Snowden, January 1851), pp. 4–8. See also Marshall Lefferts, "Telegraph and Geographical Distribution." Lefferts, and later Craig, were among the pioneers in subsequent efforts to coordinate the operations of the country's various telegraph lines to form a smooth and coherent national network, however diverse and fragmented its ownership.
18. An important technique that made the system possible was the development of manifold paper that could make up to ten copies in one writing. Reid, *Telegraph in America,* pp. 786–797; Smith, "Press as News Gatherer," p. 532; Shanks, "How We Get Our News," p. 516.
19. The classic description of natural monopolies is Frederik Zeuthen, *Problems of Monopoly and Economic Warfare* (London: Routledge, 1930). For a summary see C. E. Ferguson, *Microeconomic Theory* (Tel Aviv: Schocken Publishing House, 1975), pp. 224–265 (in Hebrew).
20. Thompson, *Wiring a Continent,* pp. 240–255.
21. O'Rielly, "Associated Press," p. 246.
22. New York and New England Telegraph Company, *Report,* 31 March 1852, Smith papers, Maine.
23. Thompson, *Wiring a Continent,* p. 194; Reid, *Telegraph in America,* p. 381.
24. Alexander Jones to Smith, 24 and 31 May 1852; John McKinney to Smith, 31 May 1852, Smith papers, Maine.
25. The new company was capitalized at $300,000. Of this highly watered stock, the Magnetic, or Morse, line received $200,000, and the Bain interests $100,000. Articles of Association of the New York and New England Telegraph Company, 1 July 1852, Smith papers, Maine. An accord between Smith and O'Rielly was imminent since their appeasement at the close of the previous year and their abortive settlement of December 1851. O'Rielly, "Associated Press," pp. 244–245; Craig, *Letter to Edson,* pp. 16–19; Thompson, *Wiring a Continent,* p. 192.

26. Quoted in John McKesson to Smith, 26 July 1852, Smith papers, Maine.
27. Smith to O'Rielly, 2 September 1852, quoted in O'Rielly, "Associated Press," p. 255.
28. *New York Courier and Enquirer,* 13 July 1852, quoted in O'Rielly, "Associated Press," p. 246. McKesson to Smith, 26 July 1852, Smith papers, Maine.
29. Craig, printed letter to Freeman M. Edson, p. 6. The letter was published as a pamphlet no later than June 1853. Pages 3–10 of this pamphlet (without a title page) are in the Smith papers, Maine. (Subsequent references will be to *Edson.*)
30. McKesson to Smith, 24 July 1852, Smith papers, Maine. The best account of the involved claims on the House line is in a "statement of facts" by one of the stockholders, appended to Craig, *Letter to Edson,* pp. 20–26. Quotation from p. 7.
31. McKesson to Smith, 24 July 1852, Smith papers, Maine.
32. Charles Rogers, the leading spirit of the opposition association, soon informed Smith that "the Bain portion of the new line [is] determined to drive me and my papers over to the tender mercies of D. H. Craig," "a man whom you love no better than me." When wires were down or business was heavy, "Craig gets his news through, but Abbot is denied a turn at the wires . . . He [Abbot] is absolutely ashamed to go in the office." Rogers had his own embarrassments: "I am absolutely ashamed to own," he wrote Smith, "that I have been all my telegraphic life in favor of a line which owning two thirds of the stock—allows me to be so mercilessly beaten." Rogers to Smith, 6 September 1852; 11 September 1852 (two letters); 14 July 1852, Smith papers, Maine. Smith may have made some promises, or else Rogers found it politic to represent Smith as promising his association relief. The Boston evening papers began insinuating that after the end of the year during which the Bain people managed the company, their news interests would be taken care of by Smith. Smith was called on to allay the fears of an alarmed AP. McKinney to Smith, 15 July 1852; McKesson to Smith, 24 July 1852; George N. Lawrence to Smith, 26 July 1852; McKesson to Smith, 26 July 1852; McKesson to Smith, 22 August 1852, Smith papers, Maine.
33. Craig, *Letter to Edson,* p. 15.
34. Craig, *Edson,* pp. 5–6. O'Rielly subsequently claimed that Craig, with whom the Union line conducted the negotiations, failed to inform members of the association of the five-year offer, hoping to present his arrangement with the House line as the only reasonable solution. Although Gerard Hallock, editor of the *Journal of Commerce* and president of the association, made such a claim, it appears it was his way of

dodging O'Rielly's complaints and that the association reached its decision unanimously. O'Rielly, "Associated Press," pp. 254–256; but see Craig, *Edson,* p. 6, and *Letter to Edson,* p. 17.

35. *New York Sun,* 6 July 1849.
36. Townsend to Cornell and Speed, 29 August 1851, Ezra Cornell papers, Cornell University Library; Schwarzlose, *Newsbrokers,* vol. 1, pp. 192–193.
37. Alfred D. Chandler, *The Visible Hand: The Managerial Revolution in American Business* (Cambridge: Harvard University Press, 1977), pp. 9–10, 28–48.
38. Harbor News Association agreement, 1 June 1848, article 5, Moses S. Beach papers.
39. Henry O'Rielly to Gerard Hallock, 7 and 22 September 1852, quoted in O'Rielly, "Associated Press," pp. 255–257. Daniel H. Craig, *A Reply to the Falsehoods of the Executive Committee,* pamphlet (New York, 7 July 1860), p. 9; D. H. Craig, "Reply of the General Agent of the Press to the Executive Committee," *New York Herald,* 23 April 1860. The Union line offered the association a contract for five years or even longer. It was willing to let the association set its own price for the dispatches. Its own proposition amounted to the charge of a quarter of a cent a word. The Associated Press ultimately contracted with the House line for three cents a word.
40. Craig, "Reply of the General Agent"; Marshall Lefferts to the Board of Directors, Union Telegraph Company, 14 December 1852, and Treasurer's Report, 21 December 1852, New York and New England Union Telegraph Company, *Proceedings,* pp. 41, 56; Craig, *Testimony,* p. 1268. The dates given by Reid, *Telegraph in America,* pp. 410–411, for the decision of the AP and the purchase of the House line are inaccurate. Consequently, Reid and those who accepted his authority did not realize the significance of the purchase for press business. It may have been that Craig did not actually invest money in the line but was given one-third of the ownership in consideration of his bringing the AP's business over to the line. Russell, *Remarks,* p. 65.
41. Quoted in Craig, *Edson,* p. 6.
42. McKesson to Smith, 13 February 1858, Smith papers, Maine.
43. Lefferts to Craig, 26 November 1852, quoted in Union Telegraph Company, *Proceedings,* pp. 57–58; O'Rielly, "Associated Press"; Craig, *Letter to Edson,* p. 9.
44. Craig to Lefferts, 29 November 1852; Lefferts to Craig, 29 November 1852; Craig to Lefferts, 29 November 1852, in Union Telegraph Company, *Proceedings,* pp. 58–61; Craig, "Reply of the General Agent"; Craig, *Reply to the Falsehoods,* pp. 4–7.
45. On the relations between the Nova Scotia and New Brunswick compa-

nies see the extensive correspondence between officials of those lines, and their correspondence with Eddy of the Maine line, Lefferts of the Boston line, and Craig, in Nova Scotia Telegraph Correspondence, RG 31, series 120, vol. 10, Public Archives of Nova Scotia. Quotation is from Craig, *To the Members.* An insightful analysis of this issue is contained in the 1850 correspondence between Lefferts and Craig. Craig to Lefferts, 11 October 1850; Lefferts to Craig, 26 October 1850; Craig to Lefferts, 5 December 1850, quoted in Marshall Lefferts, *Address to the Stockholders of the New York and New England Telegraph Company,* pamphlet (New York: Snowden, January 1851), pp. 4–8.

46. Craig, *Testimony,* p. 1268.
47. Reid, *Telegraph in America,* p. 423.
48. Ibid., p. 414. See also Craig, *Reply to the Falsehoods,* p. 10.
49. Soon after its purchase, Craig's line became subject to much solicitation. Because of McKinney's energetic management and the patronage of the AP, the line got more than half of the rapidly growing business between Boston and New York, and it paid unprecedented dividends, at times as high as 50 percent yearly on the investment. Craig, *Reply to the Falsehoods,* p. 10. Yet the real intention behind the high bids was to re-sell the line to the Union company for a fabulous price. Lefferts to Smith, 10 June 1853, Smith papers, New York; Saddler to Smith, 26 July 1853 and 18 August 1853, Smith papers, Maine; Craig, *Edson,* pp. 7–9; Craig, *Letter to Edson,* pp. 3–13. The expanded business between the cities made many realize that exclusiveness on the route had great value. "The route to Boston is cheap," suggested one of the directors of the Union line, "if we can maintain the monopoly at $500,000." McKesson to Smith, 26 July 1852, Smith papers, Maine.
50. Vary T. Coates and Bernard Finn, *A Retrospective Technology Assessment: Submarine Telegraphy* (San Francisco: San Francisco Press, 1979), pp. 7–24.
51. The Maine company purchased Smith's Boston and Portland line in September 1853. Reid, *Telegraph in America,* p. 395. Craig controlled with his partners the lines of the East and West, Troy and Boston, and the Nahant and Boston companies (p. 407). He was also part owner in the cables to Cape Cod, to Sandy Hook, and across New York's East River, and later, the telegraph line to Belle Isle.
52. Reid, *Telegraph in America,* pp. 405–406; Thompson, *Wiring a Continent,* pp. 303–305; see also Craig to Field, 13 December 1858, and Hughes to Field, 22 November 1858, Field papers. Craig's endowment to the new company included also a grand idea, in embryonic form, that had the potential of revolutionizing the state of American telegraphy. This was a petition to Congress, signed by Craig and Eddy, among others,

proposing to establish an overland telegraph line to the Pacific Coast. The New York group, however, had as its goal London rather than San Francisco, and the enterprise was allowed to die quietly.

53. Craig's stock in the company was entrusted to Hiram Hyde, the Truro transportation entrepreneur who ran Craig's Halifax Express and had built segments of the Nova Scotia telegraph line. Indenture between William Taylor and Hiram Hyde, 10 July 1849; Hyde to Joseph Howe, 9 July 1849; Taylor to Howe, 7 July 1849, RG 1, vol. 442, Public Archives of Nova Scotia; Regan, "Inception"; Reid, *Telegraph in America,* p. 405.
54. For a summary, see Reid, *Telegraph in America,* pp. 404, 414–415; Thompson, *Wiring a Continent,* pp. 301–306.
55. Craig to John Caton, 29 December 1856; see also Craig to Caton, 16 February 1855, Caton papers, Library of Congress. For other antecedents of the arrangement see Craig to Cornell, 22 January 1856; 11 February 1856; 19 February 1856; Craig to Clerk of Western Union Telegraph Office, 21 February 1856, and cover note from Craig to Cornell, Cornell papers.
56. See, e.g., Field to Smith, 15 May 1855; Field to Smith, 19 June 1855; Smith to Field, 21 June 1855; Field to Smith, 25 June 1855, Smith papers, Maine. Magnetic and Union Telegraph Companies: *Memorial of the Magnetic Telegraph Company and the New England Union Telegraph Company to the Congress of the United States,* pamphlet (10 March 1858); *British Telegraph Monopoly,* pamphlet (1858); *Memorial to the Senate and House of Representatives of the United States of the Magnetic and New England Union Companies, in Reply to the Remonstrance of the American Telegraph Company,* pamphlet (1858). American Telegraph Company, *Answer and Remonstrance of the American Telegraph Company to the Memorial of the Magnetic and New England Union Telegraph Companies,* pamphlet (20 April 1858).
57. Agreement, General News Association of the City of New York, 21 October 1856, Moses S. Beach papers.
58. Russell, *Remarks,* p. 50.
59. Smith, *Boston Daily Courier,* 10 February 1859.
60. Zenus Barnum, Hiram O. Alden, and Robert W. Russell, *The American Telegraph Company and the New York Associated Press* (New York, 22 June 1860), p. 20.
61. Craig, described in Reid, *Telegraph in America,* pp. 794–795; Craig, *Reply to the Falsehoods,* p. 25.
62. Craig, "Sworn Statement."
63. John Mullaly, *A Trip to Newfoundland: Its Scenery and Fisheries, with an Account of the Laying of the Submarine Telegraph Cable* (New York: T. W.

Strong, 1855); Reid, *Telegraph in America,* p. 402; Thompson, *Wiring a Continent,* p. 306.

64. Reid, *Telegraph in America,* p. 402. Reid estimated the price of the Newfoundland line at one million dollars, whereas Russell estimated it at "seven and eight hundred thousand dollars," *Remarks,* p. 71. Russell made there a strong statement about the uselessness of the Newfoundland line. G. G. Hubbard considered the investment of the New York group in the line $800,000. *Boston Evening Transcript,* 16 February 1859.
65. *London Times,* 30 August 1858; and see Ben Edwards Grey, *Foreign Mails—Ocean Steamers* (New York: William C. Bryant and Co.), p. 4.
66. *New York Times,* 27 August 1857 and 25 April 1857; Charles F. Briggs and Augustus Maverick, *The Story of the Telegraph and a History of the Great Atlantic Cable* (New York: Rudd and Charelton, 1858); John Mullaly, *The Ocean Telegraph* (New York, 1859); Thompson, *Wiring a Continent,* p. 319. The standard nineteenth-century accounts of ocean telegraphy are Henry M. Field, *The Story of the Atlantic Telegraph* (New York: Charles Scribner's Sons, 1893), which includes a long section on the attempts of 1858, and Isabella Field Judson, *Cyrus W. Field: His Life and Work* (New York: Harper and Brothers, 1896).
67. Kendall and Smith entered into a close alliance in opposition to the American Telegraph Company, and besides their attempts to block government aid to the cable enterprise, they strengthened their position considerably by securing a lease of the line from Washington to New Orleans. Kendall to Smith, 6 January 1857 and 31 October 1857, Smith papers, New York; Smith to Kendall, 30 December 1857; McKesson to Smith, 13 February 1858; Smith to McKesson, 19 February 1858, Smith papers, Maine. *New York Times,* 5 April 1858, 12 June 1858; Magnetic and Union Telegraph Companies: *Memorial; British Telegraph Monopoly; Memorial in Reply;* American Telegraph Company, *Answer and Remonstrance.*
68. *New York Times,* 18 August 1858; quoted also in Charles F. Briggs and Augustus Maverick, *The Story of the Telegraph and the History of the Great Atlantic Cable* (New York: Rudd and Charelton, 1858), pp. 246–247.
69. Quoted in Judson, *Field,* pp. 20–22; Mullaly, *Ocean Telegraph,* pp. 281; *New York Times,* 6 August 1858. There is a glaring discrepancy between the report of the landing and the reports issued by the Atlantic Cable Company in London. Field stated on 5 August that he had received the signals for the cable's landing in Valentia, Ireland, on 4 August, whereas the English reports stated that they landed their end of the cable on 5 August. *London Times,* 11 August 1858.
70. Quoted in the *New York Times,* 9 August 1858. The most comprehensive account of the public response to the cable is provided in Philip B.

McDonald, *A Saga of the Seas: The Story of Cyrus W. Field and the Laying of the First Atlantic Cable* (New York: Wilson-Erickson, 1937), ch. 5.

71. "There was a deep-seated feeling in the community that the cable had never transmitted a letter." Massachusetts Senate, *Debate,* p. 23; Harlow, *Old Wires,* pp. 240–242; Prescott, *History Theory and Practice of the Electric Telegraph* (Boston: Tickenor and Fields, 1860), pp. 206–208; Walter P. Phillips, *Sketches Old and New* (New York: George Munro's Sons, 1897), p. 194; see also *Boston Courier,* 3, 12, 14, 15, and 21 February 1859, and *New York Herald,* 3 and 4 August 1866.
72. Harlow, *Old Wires,* p. 242; *Boston Courier,* 3, 12, 14, 15, and 21 February 1959.
73. According to Prescott, De Sauty surfaced again in 1860 as the electrician of the cable between Malta and Sicily. *History Theory and Practice,* pp. 226–227.
74. "De Sauty: An Electro-Chemical Eclogue," in *The Professor at the Breakfast Table* (Boston: Houghton Mifflin Co., 1893), pp. 26–28.
75. Even the president was not sure that the awkward message tendered to him by the AP was not a fabrication. He was, however, assured by the superintendent of the Newfoundland telegraph that the dispatch was the same one received from Trinity Bay and transmitted over his line; hence, it was patently the queen's message. Surprisingly, this faulty validation of the bungled dispatch convinced the public. *New York Times,* 16 August 1858. The president, however, still "properly exercised as to its truth, not wishing to show his gallantry in the absence of a proper invitation to do so," called the Washington agent of the AP into his office. He asked a few questions, "wanting to be certain that the Queen had sent him the message." Lawrence Gobright, the agent, "satisfied him that he ran no risk in responding to the lady." The president presented the agent with a gracious reply to the queen and allowed him to make it public. Yet, instead of depositing the official reply in the "public archives" and having Gobright take the copy for transmission, the president insisted, against the opinion of his advisors, on having the agent take the original. President Buchanan, perhaps not yet fully convinced of the authenticity of the cable, may have feared going down in history as the victim of a hoax. Lawrence A. Gobright, *Recollection of Men and Things at Washington* (Philadelphia: Claxton, Remsen, and Haffelfinger, 1869), pp. 190–191. A week earlier, the president, then on vacation at Bedford Springs, had promised to authorize the AP to make the queen's and his own messages public. *New York Times,* 10 August 1858.
76. The message was published in the United States in two installments, on 16 and 17 August, owing to faulty transmission. Yet the English Company claimed that the message was transmitted complete in

sixty-seven minutes from Ireland to Newfoundland, and was repeated in its entirety from Trinity Bay to Valentia. *London Times,* 18 August 1858. See also correspondence between Seward and Whitehouse, 16 and 19 August 1858, in *Recent Correspondence between Mr. Wildman Whitehouse and the Atlantic Telegraph Co.* (London, 1858), which implies that the messages were not transmitted on the announced dates.

77. For early complaints of the procedures for informing the public of the cable operations, see *New York Times,* 14, 16, and especially 18 August 1858.
78. I have surveyed the files of the *New York Times* and of the *London Times* for August 1858. Of all the dispatches reported by the NYAP and the *London Times* as coming over the cable, only one broke a news event that occurred less than a week before it was made public as a cable dispatch. That one case was the announcement by the *London Times,* on 25 August 1858, that "Mrs. Eddy [*sic*], the first and best telegrapher in the States, died today." The death of James Eddy, one of the leaders of the American Telegraph Company, was reported to have occurred two days before, on 23 August. Yet the item was published in England and did not relieve American skeptics. Moreover, since Eddy died while on a business trip to Vermont on behalf of the American Telegraph Company, few besides the American Telegraph Company personnel could have known the exact date of his death.
79. *Boston Courier,* 3 February 1859. See also 12, 14, and 15 February 1859.
80. Yet in a suit that was brought against the cable company in England for fraud, Lieutenant Francis Higginson, an eyewitness to the operation, claimed that the cable was lost while being spliced; hence it never could have carried a current. *New York Times,* 27 October 1858.
81. *New York Times,* 28 August 1858.
82. Ibid., 30 August 1858.
83. One of the exceptions was the *Boston Courier,* which published a detailed expose on 3 February 1859. Subsequently the newspaper published a series of responses and rebuttals, on 12, 14, and 15 February 1859. The *Courier* quoted on 21 February 1859 the observation of the *New York Saturday Press* (one of the few New York newspapers not affiliated with the AP) that "the most suspicious thing in the matter is" that Observer's article "has been carefully excluded from the New York press." The occasion for the publication of the article in the *Courier* was that newspaper's campaign for an alternative telegraph line to Halifax, publicized as an effort to limit the powers of the NYAP. The other exception was the *New York Herald,* 3 August 1866 (see also 4 August 1866). The *Herald* was attempting at the time to establish a

foreign news service in competition with, and independent of, the NYAP.

84. The intention of the cable entrepreneurs to unload their stock is indicated in George Peabody to Peter Cooper, 3 [?] August 1858. Cooper, Hewitt, and Company papers, Library of Congress.
85. See, e.g., John T. Smith to F. O. J. Smith, 18 August 1858, Smith papers, Maine.
86. The bill, as reported by the Committee on Mercantile Affairs and Insurance, may be found in Massachusetts, Senate, *Legislative Documents,* March 1859, no. 109. For a summary of the opposition stand see the minority report appended to the bill (pp. 11–14). Benjamin F. Butler, the shrewdest critic of the bill, pointed out that the capitalization of the company seeking incorporation was unlimited, thus enabling it to reach any destination around the globe. Moreover, in contradiction to the supplement to the Massachusetts Telegraph Act that made it compulsory for companies to receive and deliver messages to any other company, the new company legislated itself the right to make exclusive connections with any lines it chose to. The Committee on Mercantile Affairs and Insurance that reported the bill found it necessary to amend it under the force of Butler's arguments. Massachusetts Senate, *Debate,* pp. 17–21. Curiously, the *Boston Courier,* an ardent supporter of the bill, reported Butler's brilliant criticisms thus: "His chief objection to the measure was a personal dislike to Mr. F. N. Gisborne"; and it added further frivolous details in reporting Butler's objection to the bill. *Boston Courier,* 24 March 1859.
87. Craig to Thomas Bolton, 7 January 1859 and 6 April 1859; W. H. Wiswell to W. H. Abel, 18 May 1859; and other letters in Nova Scotia Telegraph Company, *To the Public,* pamphlet (1859).
88. *Memorial of the Magnetic; Answer and Remonstrance; Memorial in Reply to the Remonstrance; British Telegraph Monopoly;* Thompson, *Wiring a Continent,* pp. 320–321.
89. "Meeting in the Merchant's Exchange Reading Room," *Boston Courier,* 14 January 1859.
90. Quotations from the *Boston Courier,* 26 March 1859; and Massachusetts, Senate, *Debate,* p. 35.
91. Massachusetts, Senate, *Debate,* pp. 27, 34.
92. The bill was lost on the third reading, after several senators reversed their vote and several erstwhile supporters failed to vote. Supporters argued that unscrupulous lobbyists were responsible for the defeat of the bill, and they insinuated that the result of the vote was achieved through bribery. *Boston Courier,* 30 and 31 March 1859. Matthew D. Field, who went to Boston to lobby the legislature on behalf of the American Telegraph Company (the company was not granted a

standing in the committee hearings), expected his expenses to amount to $50. He ultimately presented the company with a bill for expenses of $3,000. Russell, *Remarks,* pp. 11–12; see also Appendix to Massachusetts, Senate, *Debate,* pp. 45–46.

93. Massachusetts, Senate, *Debate,* p. 37.
94. Ibid., pp. 27–28.
95. *Boston Courier,* 10 February 1859.
96. Massachusetts, Senate, *Debate,* p. 41. From a cursory survey of the various votes on the bill, it appears that senators representing counties in eastern Massachusetts tended to support the bill, whereas those from western Massachusetts were opposed to it.
97. W. G. Blanchard, "Statement," *Boston Courier,* 12 February 1859; *Boston Herald, History,* p. 37. See also *Boston Evening Gazette,* 20 November 1858.
98. *Boston Advertiser,* 21 March 1859, emphasis added.
99. Massachusetts, Senate, *Debate,* p. 41.

6. "To Liberate the Newspaper Press"

1. "We intend to *liberate* 'the newspaper press' from the yoke of D. H. Craig and the New York Association, and we do not care a straw about the hostility of that association, backed up as we shall be by the press of the country at large and the commercial community." Russell, *Remarks,* p. 49.
2. Craig was trying to force the Nova Scotia Telegraph Company to sell out or lease its lines to the American. But the Nova Scotians demanded 7 percent on capital of $120,000 for the lease, a rate that Craig and his telegraph allies found too high. Craig, "Sworn Statement."
3. Craig, "Sworn Statement." For the financial condition of the company, see Statement of Receipts and Expenditures of the Nova Scotia Electric Telegraph Company for 1853, Nova Scotia, House of Assembly, *Journal* (1854), appendix 13, pp. 98–99; statement for 1854, *Journal* (1855), appendix 4, pp. 60–61. From the statement of 31 December 1852, *Journal* (1853), appendix 6, pp. 162–165, it appears that the AP business accounted for some 40 percent of the company's income in 1852. All the offices besides the one at Halifax operated at a loss and were kept open by the government as a service to the public. From a purely commercial standpoint, therefore, the steamer dispatches constituted about 60 percent of the legitimate business of the company.
4. D. H. Craig, *To the Agent of the Associated Press,* 10 May 1859, printed circular. The best sources for the affair are Craig, "To the Executive Committee," 3 June 1859, appended by a statement of the American Telegraph Company, *New York Tribune,* 4 June 1859; *New York Herald,*

4 June 1859. The statements include extensive quotations from the original correspondence between the parties to the conflict. Major parts of this original correspondence were reproduced in Nova Scotia Telegraph Company, *To the Public,* pamphlet (1859). See also *New York Times,* 4 June 1859, for a good summary of developments. Russell, *Remarks,* pp. 19–22, 57–69, includes an important account of the affair and quotes much of the correspondence between the parties.

5. Craig, "Sworn Statement."
6. James C. Cogswell, President, Nova Scotia Telegraph Company, to W. H. Abel, Secretary, American Telegraph Company, 25 May 1850, quoted in "The Telegraphic Quarrel," *New York Tribune,* 4 June 1859.
7. Gisborne had initiated the Newfoundland telegraph line, but in 1857 he fell out with Field and his group, as well as with Craig, with whom he was "on terms of intimacy" since 1849. Craig, "Sworn Statement." Gisborne's career after that point became somewhat checkered. He became involved in some shady silver mining operations in Newfoundland, and after their failure, "obliged to do something besides flying kites," as Craig observed, he returned to the telegraphic arena. He entered into a grandiose cable venture, with the intention of competing with his former employer, Field. The Cape Ann-Yarmouth line was the first leg of his trans-Atlantic cable project. For details on Gisborne's career, see Joseph Howe to Gisborne, 12 February 1867 (broadside), and various other broadsides and newspaper clippings in MG 100, vol. 146, Public Archives of Nova Scotia; and Massachusetts, Senate, *Debate.* Smith became interested in the project during the course of intensive negotiations with the American Telegraph Company for the sale of all his telegraphic interests. The new project strengthened his hand in the bargaining and provided some insurance to his Union line between Boston and New York in case the negotiations fell through. *Boston Courier,* 4 and 5 February 1859.
8. The Nova Scotia Company expected to collect tolls for transmitting steamer news from Halifax to Yarmouth. From there the news would pass over Gisborne and Smith's cable directly to Cape Ann, and from Boston to New York over Smith's Union line.
9. For Johnson's appeal with A. W. Carr to the Nova Scotia authorities, see Chapter 5. For detailed accounts of the "news conspiracy," see *American Telegraph Magazine,* 1:1 (October 1852), p. 7; 1:3 (December 1852), p. 131; 1:4 (January 1853): 158–161; 1:5 (February and March 1853): 217–222. According to Craig, Johnson's subsequent career consisted of some "twenty brilliant schemes for 'smashing the Associated Press Machine.'" Upon the failure of each, Johnson returned "for a brief period to his old quarters under the shadow of the small beer shops of Philadelphia." Craig, *Reply to the Falsehoods,* p. 22.

10. As late as December 1858 Johnson was still associated with Carr in his news arrangement. The great expense involved in picking up the contract with the Nova Scotia telegraph probably made him choose a more affluent partner. Joseph Harding et al., circular to the press of Philadelphia, quoted in Russell, *Remarks,* p. 58.
11. Craig, *Reply to the Falsehoods,* p. 22.
12. Kendall to Morse, 24 August 1859, Morse papers; Craig, "Reply of the General Agent of the Press to the Executive Committee," *New York Herald,* 23 April 1860; Craig, *Reply to the Falsehoods,* pp. 16–17, 19.
13. See Chapter 5.
14. Craig, *To the Agent;* Craig, *Testimony,* p. 1279; Reid, *Telegraph in America,* pp. 369–370. Several details of Craig's testimony, given some fifteen years after the fact, appear flawed. Reid's account of the conflict is entirely inaccurate.
15. Zabriskie to Smith, 28 May 1859, Smith papers, Maine; Craig, "Sworn Statement."
16. Peter Cooper to James C. Cogswell, 4 May 1859; W. H. Abel to Cogswell, 13 May 1859; Abel to W. H. Wiswell, 17 May 1859; H. O. Alden to Cogswell, 21 May 1859. All are quoted in "The Telegraph Quarrel," *New York Tribune,* 4 June 1859.
17. Eventually, Johnson and Zabriskie sued the American for stopping their messages. The American, by then antagonistic toward the AP, announced that it was sued and "had to pay damages." American Telegraph Company, *The Telegraph and the Press,* p. 24. David Dudley Field, the foremost jurist on the board, did not believe such a verdict was merited, and his persistent inquiries made the executive committee admit that no decision was given in court and that they paid $2,962 in damages voluntarily. Russell, *Reply,* p. 20.
18. Kendall to Morse, 24 August 1859, Morse papers; Thompson, *Wiring a Continent,* pp. 327–328.
19. Craig, "Reply of the General Agent."
20. Ibid.; Craig, *Reply to the Falsehoods,* pp. 25–26; Russell, *Remarks,* pp. 31–32; Thompson, *Wiring a Continent,* pp. 327–329.
21. Agreement between the American Telegraph Company and the Magnetic Telegraph Company, 12 October 1859, quoted in American Telegraph Company, *The Telegraph and the Press,* p. 25.
22. Kendall to Morse, 24 August 1859; Thompson, *Wiring a Continent,* p. 328.
23. Craig, *Edson;* Craig, *Letter to Edson;* Reid, *Telegraph in America,* pp. 424–425.
24. Thomas Bassnett to John D. Caton, late July or early August 1860, Caton papers.
25. AP proprietors righteously lamented their neglect of one of their "first

and most imperative duties"—warning the public "against the creation of such a gigantic and unscrupulous monopoly." *New York Times,* 8 July 1860.

26. American Telegraph Company, *The Telegraph and the Press,* p. 20.
27. Russell, *Remarks,* p. 49.
28. Ibid., p. 35. The evaluation of Russell is from Reid, *Telegraph in America,* p. 424. For a more blunt description of Russell, see Bassnett to Caton, July or August 1859, Caton papers.
29. Russell, *Remarks,* pp. 22–23.
30. Craig, "Reply of the General Agent"; Craig, *Reply to the Falsehoods,* p. 14; *New York Herald,* 30 June 1859. Rosewater apparently followed Reid in an inaccurate account of the entire affair. He mistakenly attributed the revised price policy of the Newfoundland line—charging the AP according to the importance of the news—to the American Telegraph Company. *Cooperative News Gathering,* p. 95.
31. Russell, *Reply,* p. 51.
32. Ibid., p. 55.
33. *New York Times,* 3 July 1860; see also the statement of F. O. J. Smith to the stockholders of the American Telegraph Company, 29 June 1860, in the *New York Herald,* 30 June 1860.
34. The Field-Cooper-Hunt group opposed the measure, arguing that "the right to combine to procure intelligence is as sacred as the right of the stockholders of this company to associate themselves together to build and operate telegraph lines." Statement of Wilson G. Hunt, Cyrus W. Field, and Abram D. Hewitt to the stockholders of the American Telegraph Company, quoted in the *New York Herald,* 30 June 1860. Russell responded that he did not find any right—"divine, sacred or otherwise to combine to procure intelligence." Russell, *Remarks,* p. 31.
35. The issue of the prices charged the Boston press was discussed extensively in numerous pamphlets and newspaper articles. See "The Executive Committee of the American Telegraph Company to the Press," *New York Herald,* 23 April 1860; Craig, "Reply of the General Agent"; American Telegraph Company, *The Telegraph and the Press;* Craig, *Reply to the Falsehoods;* also a pamphlet issued by the executive committee as a reply to Craig's latter pamphlet, apparently entitled: *The Telegraph and the Press: A Report to the Press of the Actions of Various Committees and Resolutions Leading up to an Increase in Tariff on Press Messages* (1860). An incomplete copy is in the New York Public Library (only pp. 13–35 extant); "The Telegraph and the Press," *Springfield Republican,* undated slip in the Moses S. Beach papers, with a note in Craig's hand (and signed "DHC"), "Will not Mr. Beach give these fellows a *first rate notice?*"

36. Russell, *Remarks,* pp. 34, 38; see also *Report to the Press,* incomplete pamphlet (New York Public Library), p. 19.
37. *Report to the Press,* p. 33.
38. Amos Kendall, *Circular to the Stockholders of the American Telegraph Company,* 27 August 1860, p. 12; Craig, *Reply to the Falsehoods,* pp. 32, 34; see also *New York Times,* 31 July 1860; Hudson, *Journalism in America,* p. 614; Reid, *Telegraph in America,* p. 424; Thompson, *Wiring a Continent,* p. 339.
39. *New York Times,* 31 July 1860.
40. Craig, quoted in Kendall, *Circular,* p. 5; see also Craig's confidential letter to Kendall, quoted without permission (p. 6).
41. James Watson Webb et al., to Zenus Barnum, President, American Telegraph Company, 30 May 1859, quoted in Hunt, Field, and Hewitt, "Statement"; and in American Telegraph Company, *The Telegraph and the Press,* pp. 29–31.
42. Craig, *Reply to the Falsehoods,* p. 35.
43. Russell, *Remarks,* pp. 54–55.
44. Kendall, *Circular,* p. 4.
45. Craig, "Sworn Statement."
46. Russell obviously viewed patent rights as the only obstacle to an AP line. He explained that he timed his move as an ambush, leaving himself almost a year to crush the association before the expiration of Morse's patent would allow it to construct competing lines. Yet O'Rielly, who purportedly was the man chosen by Craig to undertake the project, assured the association that a suitable patent would be found. Besides, once a competing line was constructed, the lawsuits to prove that a slightly modified telegraph instrument constituted an infringement on one of the existing patents would probably drag on until the date when Morse's patent would expire.
47. Russell, *Remarks,* p. 37.
48. The press between Boston and Washington paid about $50,000 annually for messages from New York. The New York papers paid about $100,000 to the American Telegraph Company. Although part of this sum was for service rendered east of Boston and south of Washington, the press east and south of the Boston-Washington axis paid more than $100,000 annually for press reports, and part of this sum was for transmission within the Boston-Washington zone. Hence, the conservative estimate in the text. Craig, *Reply to the Falsehoods,* pp. 12, 24.
49. Estimated yearly expenses per wire-mile were calculated by dividing the annual expenses of the American Telegraph Company—$735,008 (Address of Cyrus W. Field to the stockholders of the American Telegraph Company, 29 June 1860, in the *New York Herald,* 30 June 1860), by the number of wire-miles the company operated—13,015. Robert

W. Russell, *Statistics, etc., of the American Telegraph Company,* pamphlet (New York, 20 August 1860), p. 15.

50. Quoted in the statement of Charles Hale (proprietor of the *Boston Daily Advertiser*) to the stockholders of the American Telegraph Company, 29 June 1860, in the *New York Herald,* 30 June 1860.
51. Craig, *Reply to the Falsehoods,* pp. 12, 24.
52. Russell, *Statistics, etc.,* pp. 9–11; Field, "Statement."
53. Statement of F. O. J. Smith to the stockholders of the American Telegraph Company, 29 June 1860, in the *New York Herald,* 30 June 1860.
54. Reid, *Telegraph in America,* pp. 427–432; Thompson, *Wiring a Continent,* pp. 310–330, 339–342. The parties to the contract, entered into on 10 August 1857, were the American Telegraph Company, operating along the eastern seaboard; the New York, Albany, and Buffalo Telegraph Company; the Atlantic and Ohio Telegraph Company, operating in Pennsylvania and the middle-Atlantic states; the Western Union Telegraph Company, then controlling lines mainly in Ohio, Indiana, Michigan, and Wisconsin; the New Orleans and Ohio Telegraph Company, operating in the Southwest; and the Illinois and Mississippi Telegraph Company, which served the northwestern states.
55. Hiram Sibley to John D. Caton, 9 April 1860, Caton papers.
56. Sibley to Caton, 11 March 1860 and 9 April 1860; Bassnett to Caton, July or August 1860, Caton papers; Craig, *Reply to the Falsehoods,* pp. 31, 35; Russell, *Remarks,* pp. 22–23, 36–37.
57. Thompson, *Wiring a Continent,* pp. 341–342. Thompson and those who followed him did not realize that the leasing of the Newfoundland line was what settled the issue.
58. *New York Times,* 5 October 1860, *New York Herald,* 5 October 1860; Craig, *Testimony,* p. 1268; Hudson, *Journalism in America,* p. 614. Reid, *Telegraph in America,* p. 369.
59. At the time, Craig was apparently venturing on a telegraph line to Belle Isle, where his men could intercept the steamers of the Quebec-Galway line.
60. The *Herald* stated that in the course of 1860 "the news yacht has been enabled to intercept some thirty or forty steamers, but in the future we shall be able to do better than that." *New York Herald,* 5 October 1860. These hopes never materialized. Steamers took the northern route only eight months a year, and weather conditions made it impossible to intercept many of them. Hudson stated that the association intercepted thirteen outbound steamers in 1859; thirty-one in 1860; thirty-four in 1861; and forty-six in 1862. *Journalism in America,* p. 616. Craig stated that thirty-four reports were obtained in 1861, and sixty in 1862. New York Associated Press, *Annual Report* [1863], p. 4. See also Coats and Finn, *Retrospective Technology Assessment,* p. 79.

61. *New York Times,* 5 October 1860.
62. Hudson, *Journalism in America,* p. 614. According to Craig's description of the contract (*Testimony,* p. 1279), it would appear that it was similar to the contract of 1865 between the American Telegraph Company and the New York Associated Press. William Henry Smith papers, Indiana Historical Society, Indianapolis.
63. Henry Villard, *Memoirs of Henry Villard, Journalist and Financier,* 2 vols. (Boston, 1904), vol. 1, pp. 140–152, but see editorial note on p. 140, stating that Villard was actually employed by the *New York Herald.* The AP circulated the story because of the regulation that member newspapers had to share their specials. See also Oliver Gramling, *AP: The Story of the News* (New York: Farrar and Rinehart, 1940), p. 39; Robert S. Harper, *Lincoln and the Press* (New York: McGraw-Hill, 1951), pp. 81–83; J. Cutler Andrews, *The North Reports the Civil War* (Pittsburgh: University of Pittsburgh Press, 1955), pp. 77–79; Harold A. Williams, *The Baltimore Sun, 1837–1987* (Baltimore: The Johns Hopkins University Press, 1987), pp. 44–45; Benjamin P. Thomas, *Abraham Lincoln* (New York: The Modern Library, 1968), pp. 239–244.
64. Gobright, *Recollections,* pp. 348–350; John Tebbel and Sarah Miles Watts, *The Press and the Presidency* (New York: Oxford University Press, 1985), p. 201. See also Arthur F. Loux, "The Mystery of the Telegraph Interruption," *Lincoln Herald* 81 (Winter 1979): 234–237. One account puts the number of Washington reporters during the Civil War at fifty. Henry Van Ness Boynton, "The Press and Public Men," *Century Magazine* 42 (October 1891), p. 853.
65. Shanks, "How We Get Our News," p. 518.
66. Gramling, *AP,* p. 42; Craig, *Annual Report of the General Agent* (New York: 1 January 1862); Andrews, *The North Reports,* pp. 31–32. Andrews compiled the most complete catalogue of war correspondents, and only 13 of them were connected at any time with the AP.
67. Villard, *Memoirs,* vol. 1, pp. 153–154, vol. 2, p. 267; Andrews, *The North Reports,* pp. 31–32; Andrews, *Pittsburgh's Post Gazette* (Boston: Chapman and Grimes, 1936), p. 158; Williams, *Baltimore Sun,* p. 41; Chamberlin, *Boston Transcript,* ch. 10. A good discussion of Villard's operation is in Schwarzlose, *Newsbrokers,* pp. 254–258.
68. From data in a log book of the Washington office of the House line for 1861, there appear great variations in the use of the wires between the AP members. The *Herald* was by far the largest user; the *Times* was the second largest user; then came the *Tribune;* and the *World* was a distant fourth. Entries for the AP are very irregular, and they apparently were not registered by the same standard as the individual newspapers. Donald McNicol Telegraph Collection, New York Public Library.

69. Craig put the cash revenue of the association from the outside press at $57,000, and he considered it "nearly one half" of all expenses. The figure presented in the text represents Craig's estimates for a year under recent arrangements. Craig, *Annual Report 1862.* Of this sum, $18,000 was paid for the transmission of steamer reports from Cape Race, and further funds were necessary to maintain the staff there. George Prescott, writing in 1860, estimated the annual expenses of the New York members of the AP for telegraphic reports at about $100,000. George B. Prescott, *History, Theory, and Practice* p. 33. (This figure is confirmed by Craig's accounting.) Considering that the outside press contributed a significant amount of these expenses, it appears that there was a decline in the costs of running the service from 1859 to 1862. The estimate for annual expenditure of the New York dailies for war news is from Henry Villard, "Army Correspondence," *Nation* 1 (August 1865), p. 146, quoted in Andrews, *The North Reports,* p. 20.
70. Quotation from the *Philadelphia Public Ledger,* 2 September 1856, in Elwyn Burns Robinson, "The Dynamics of American Journalism from 1787 to 1865," *The Pennsylvania Magazine of History and Biography* 61 (October 1937), p. 440.
71. Parton, "New York Herald," p. 403.
72. Ibid., p. 410. Circulation of New York dailies increased by more than 40 percent during the war years. Lee, *Daily Newspaper,* table 17, p. 731. Nationally, newspaper output in the 1860's increased by 63 percent, whereas the population expanded by only 22 percent. Dill, *Growth of Newspapers,* table 1, p. 11. For general statements of the effect of war in increasing circulation, see, e.g., Mott, *American Journalism,* p. 329; Andrews, *The North Reports,* pp. 32–34; Edward Arden, "The Evolution of the Newspaper," *Chautaquan* 29 (May 1899): 152–154.
73. Apparently, one way that members tried to evade the requirement of "sending around" their news was by demanding that parties interested in particular items share the price of the entire budget of the day's war news. Article 15 of the rules required a newspaper taking a "particular matter" from another to pay for the entire dispatch, not only its "important points." In 1861 the association had to rule that "news of movements connected with the war, in different sections of the country, does not constitute 'one particular matter.'" General News Association of the City of New York, *Rules* (New York, 1874), p. 8 (note).
74. The quotation and the general concept of the news net are from Tuchman, *Making News,* pp. 15–38 (quotation, p. 21); see also Melville E. Stone, "The Associated Press: News Gathering as a Business," *The Century Magazine* 70 (June 1905), p. 305; and Francis A. Richardson,

"Recollections of a Washington Newspaper Correspondent," *Records of the Columbia Historical Society* 6 (1903): 34–35.

75. The Press Association of the Confederate States of America, in contrast, served as a full-fledged primary news-gathering system. It appears that its cooperation with Confederate authorities was minimal, and that the association had serious problems with censorship. Press Association of the Confederate States of America, *Minutes, Rules, and Report of the Superintendent* (Griffin, Georgia: Hill and Swayze's Printing House, 1863); and *First Annual Meeting: Minutes of the Board of Directors, and Quarterly Report of the Superintendent* (Montgomery, Alabama [?]: Memphis Appeal Job Printing Establishment, 1864). For a summary, see Quintus C. Wilson, "Confederate Press Association: A Pioneer News Agency," *Journalism Quarterly* 26 (June 1849): 160–166, and J. Cutler Andrews, *The South Reports the Civil War* (Princeton: Princeton University Press, 1970). For a general discussion see David Donald, "Died of Democracy," in *Why the North Won the Civil War*, ed. David Donald (New York: The Macmillan Company, 1968), pp. 79–90.
76. See, e.g., Adam S. Hill to Sidney Howard Gay, 13 June 1962, Sidney Howard Gay papers, Columbia University, New York, New York.
77. Quotation from Gobright, *Recollections*, pp. 320–321.
78. U.S. Congress, House, *Telegraph Censorship*, 37th Cong., 2d sess., H. Rept. 64 (1862), p. 12. For examples of the great significance of transportation in war correspondence, see Andrews, *The North Reports*, pp. 214, 334, 368, 371.
79. Quoted in Robert S. Harper, *Lincoln and the Press* (New York: McGraw-Hill, 1951), p. 299.
80. "News releases from the various government departments were to be cleared through the Washington agent of the AP, who, in turn, distributed such items simultaneously to the various out-of-town newspapers." Andrews, *The North Reports*, pp. 49–50. Quotation is from Richardson, *Secret Service*, p. 258.
81. David M. Potter, *The Impending Crisis, 1848–1861* (New York: Harper Torchbooks, 1976), pp. 555–583, quotation from p. 579. For a summary, see J. G. Randall and David Donald, *The Civil War and Reconstruction* (Boston: D. C. Heath and Company, 1961), pp. 169–175.
82. Potter, *Impending Crisis*, p. 579.
83. Gobright, *Recollections*, pp. 312–314; see also Gobright to John Nicolay, 1 December 1862, Lincoln papers.
84. For examples of the AP supplying the president with information, see Craig to E. S. Sanford, 5 November 1862; Craig to Lincoln, 8 March 1864, Lincoln papers.
85. David Donald, *Lincoln Reconsidered* (New York: Vintage Books, 1961), pp. 3–18, quotation from p. 4.

86. "Telegraph Censorship," unpublished hearings of the House Judiciary Committee, 37th Cong., Records of the U.S. House of Representatives, RG 233, National Archives, Washington, D.C. During the tenure of Simon Cameron as secretary of war, Wilkeson himself had enjoyed the privilege of serving as the administration's semi-official spokesman, and his dispatches to the *Tribune* were sent uncensored. Testimony of H. Emmons Thayer [government censor], "Telegraph Censorship."
87. Re-examination of Samuel Wilkeson, and re-examination of Adam S. Hill, "Telegraph Censorship."
88. The administration's dependence on the AP was especially marked due to Lincoln's decision not to maintain a Washington newspaper as an organ. The establishment of the Government Printing Office in 1860, which curtailed patronage to the press, had an important relation to that decision. Culver Smith, *The Press, Politics, and Patronage: The American Government's Use of Newspapers, 1789–1875* (Athens: University of Georgia Press, 1977), pp. 219–234; Frederick B. Marbut, "Decline of the Official Press in Washington," *Journalism Quarterly* 33 (Summer 1956): 335–341.
89. On one occasion, when a *Tribune* correspondent brought the first news of an important battle to the president and his officers, Lincoln requested the newspaper to allow him to give the news to the AP. Emmet Crozier, *Yankee Reporters, 1861–1865* (New York: Oxford University Press, 1956), p. 391.
90. U.S. Cong., House, *Telegraph Censorship,* H. Rept. 64, 37th Cong., 2d sess., 1862, pp. 1, 12.
91. On the general problem of military censorship during the Civil War, see James G. Randall, "The Newspaper Problem in Its Bearing upon Military Secrecy during the Civil War," *American Historical Review* 23 (January 1918): 303–323; and Randall, *Constitutional Problems under Lincoln* (Urbana: University of Illinois Press, 1951), pp. 477–510. The seminal work on the First Amendment is Leonard W. Levy, *Legacy of Suppression* (Cambridge, Mass.: Harvard University Press, 1960), more recently revised as *Emergence of a Free Press* (New York: Oxford University Press, 1985). For a summary list of other views, see David P. Nord, "The Children of Isaiah Thomas: Notes on the Historiography of Journalism and of the Book in America," *Occasional Papers in the History of the Book in American Culture,* no. 1 (1987): 27–28.
92. *Telegraph Censorship,* pp. 2–3.
93. Examination of H. Emmons Thayer [government censor], "Telegraph Censorship."
94. Examination of Lawrence Gobright, ibid.

95. Ben: Perley Poore, "Washington News," *Harper's New Monthly Magazine* 48 (January 1874): 225–236.
96. Wilkeson examination and re-examination, "Telegraph Censorship."
97. Examination of Samuel Wilkeson, ibid. Wilkeson also pointed out that the alternative to collaboration with the wire service would have been disastrous to government interests as the government could not possibly escape charges of favoritism.
98. Examination of H. Emmons Thayer [government censor], ibid.
99. The arrangement through which the AP received government dispatches relating to military affairs was somewhat altered in 1864. Stanton prepared a news summary in Washington and transmitted it to Dix in New York. The latter made Stanton's dispatches available to the AP. See E. S. Sanford to Stanton, 11 May 1864, *Telegrams Collected by the Office of the Secretary of War, 1861–1882,* RG 107, Records of the Office of the Secretary of War, M-473, National Archives; Schwarzlose, *Newsbrokers,* vol. 1, p. 244.
100. The earliest comprehensive summary of the event is in NYAP Editors and Publishers to Abraham Lincoln, 19 May 1864, received 2:30 P.M. Lincoln papers; a copy can also be found in the Manton Marble papers. The best summary of the affair is an account published by the *Journal of Commerce* on 26 March 1870; it is quoted in full in Hudson, *Journalism in America,* pp. 374–376. Detailed accounts were also published in the New York dailies in the days after the affair, notably the *Tribune* and the *Times* of 19 May 1864; and in the *World* and *Journal* on 23 May 1864. See also Manton Marble, Letter to Abraham Lincoln (printed circular, 1867), Manton Marble papers. Official correspondence relating to the bogus proclamation may be found in *Telegrams Collected;* some of these dispatches were reproduced in U.S. Department of War, *The War of the Rebellion: A Compilation of the Official Records of the Union and Confederate Armies* (Washington, D.C.: Government Printing Office, 1880–1901), ser. 3, vol. 4, pp. 386–395. Discussions of the affair in the secondary literature are numerous; among the best are Schwarzlose, *Newsbrokers,* vol. 1, pp. 251–254, 256–258; Harper, *Lincoln and the Press,* pp. 289–303; and Starr, *Bohemian Brigade,* 315–320. An important account by a contemporary is David Homer Bates, *Lincoln in the Telegraph Office* (New York: The Century Co., 1907), pp. 228–243. See also Mary Cortona Phelan, "Manton Marble of the *New York World:* Champion of the Constitution" (Ph.D. diss., The Catholic University of America, 1957), pp. 32–33.
101. See, e.g., Craig to Bangs, 18 May 1864, Marble papers.
102. Craig to the proprietors of the *Journal of Commerce, Sun, Herald, Express, Tribune, World,* and *Times,* 18 May 1864, Manton Marble papers. See also Marble to Craig, 18 May 1864; Craig to Marble, 18 May 1864,

Marble papers. The fact that Craig requested the *World* and the *Journal* to publish a notice of the prize the next day proves that the full facts were known in New York before the orders of suppression came from Washington.

103. Stanton to Dix, 20 May 1864, sent 9:00 P.M. The editors were apparently notified of the order on 22 May. Dix to Stanton, 22 May 1864, received 1:45 P.M., *Telegrams Collected.*

104. Eckert to Stanton, received 2:15 P.M., 18 May 1864, *Telegrams Collected.* Reproduced in *War of the Rebellion,* ser. 3, vol. 4, p. 394. (Subsequent reference to reproduced documents will be made to *War of the Rebellion.*) On the "Independent," or United States Telegraph Company, see Thompson, *Wiring a Continent,* pp. 401–405; 421–423.

105. *War of the Rebellion,* pp. 391–393; and Dix to Stanton, 18 May 1864, received 5:40 P.M.; J. H. Foster to Stanton, 18 May 1864, received 7:25 P.M.; Caldwallader to Stanton, 18 May 1864, received 7:50 P.M., and 19 May, received 12:00 noon; Bowford to Stanton, 18 May 1864, received 11:35 P.M.; Stanton to Caldwallader and Foster, 18 May 1864, sent 11:30 P.M.; Stanton to Wallace, 18 May 1864, sent 4:30 P.M., and 19 May 1864; Stanton to Foster, 18 May 1864, sent 4:30 P.M.; *Telegrams Collected.* General Wallace, in Baltimore, was ordered to seize copies of the New York dailies carrying the bogus proclamation reaching that city by mail or express. Stanton to Wallace, 18 May 1864, sent 4:40 P.M., *Telegrams Collected.*

106. Bates, *Telegraph Office,* p. 242; Joseph Logsdon, *Horace White, Nineteenth-Century Liberal* (Westport, Conn.: Greenwood Publishing Corporation, 1971), pp. 96–98; Schwarzlose, *Newsbrokers,* vol. 1, pp. 256–258. It may have been that the Independent Telegraph Company was in the process of organizing a wire service of its own. See Francis O. J. Smith to the Executive Committee of the U.S. Telegraph Company, 16 August 1864, Smith papers, Maine; and Reid, *Telegraph in America,* p. 410.

107. Abraham Lincoln, "Order for Draft of 300,000 men," 17 May 1864, in *The Collected Works of Abraham Lincoln,* Roy P. Basler, ed. (New Brunswick: Rutgers University Press, 1953), vol. 7, p. 344; Lincoln to Salmon P. Chase, 18 May 1864 (p. 347); James Gilmore to Sidney Howard Gay, 18 May 1864, Gay papers; Stanton to Dix, 18 May 1864, sent 11:30 A.M., *Telegrams Collected;* and see telegraphic dispatch in various evening papers and late editions of morning papers, e.g., *New York Commercial Advertiser, New York Evening Post, Chicago Tribune,* 18 May 1864. See also Thurlow Weed to Stanton, 10 May 1864, Stanton papers, Manuscript Division, Library of Congress.

108. Hence, "The officer in charge of the investigation . . . reports that he is led to believe it originated in this city." Stanton to Dix, 18 May 1864,

sent 8:30 P.M.; similarly, the conviction that the proclamation was sent from Washington over the Independent's wires is reflected in B. P. Snyder to E. S. Sanford, 18 May 1864, *Telegrams Collected;* David Homer Bates diary (ms.), entry for 20 May 1864, Bates Collection, Rare Books and Special Collections, Library of Congress.

109. Bates diary, entry for 21 May 1864, Bates Collection; Bates, *Telegraph Office,* p. 243.

110. Stanton to Dix, Caldwallader, Wallace, Bowford, and Foster, 23 May 1864, sent c. 8:00 P.M., *Telegrams Collected.*

111. Siles C. Hay to the Editor of the *New York Times,* 19 May 1864, and John J. Speed to the Public, *New York Times,* 20 May 1864. Order by Edwin Stanton, 23 May 1864, Stanton papers, Library of Congress. The order includes the various conditions and stipulations set by Stanton for framing a new status for the U.S. Telegraph Company and the Inland Telegraph Company—another upstart and independent telegraph company. This order, in the form of an agreement, was signed by John J. Speed and Theodore Adams for their respective companies, in Washington.

112. On 18 May, the AP circulated a report of the expected draft of 300,000 men. See, for example, the *New York Herald* and the *New York Evening Post.* Moreover, it circulated the bogus proclamation to the western press, without contradiction, well after it was known in New York to be a forgery. Indeed, the *Chicago Tribune,* 21 May 1864, even suggested to the government that "when . . . dealing with the guilty parties . . . it will be well that it do not [*sic*] forget Mr. D. H. Craig of the New York Associated Press," the "unscrupulous news agent." For, because of his circulating the proclamation to the West, "the grief and anxiety of the country were thus a million times enhanced." See also the *Cincinnati Commercial,* 21 May 1864, quoted in the *Chicago Tribune,* 22 May 1864; and R. Yates to Lincoln, 18 May 1864, received 3:30 P.M., Lincoln papers. Craig, in a card to the western press, nevertheless insinuated that Villard's agency was responsible for the incident. Craig's card is in the *Tribune,* and see the *Cleveland Leader,* 23 May 1864. It is reproduced in full in the *Chicago Tribune,* 21 May 1864 ("The Bogus Proclamation, and Another from Mr. D. H. Craig"); and in an unidentified clipping in the Villard papers, Houghton Library. See also Craig's explanation of the AP mishap in the *Cleveland Leader,* 4 June 1864.

113. In January 1862, Charles C. Fulton, editor of the *Baltimore American* (the one pro-administration paper in Baltimore) and the AP agent there, was arrested for sending a telegram to Craig in which he offered to send a report of news from the battles on the Peninsula, "including facts obtained from Washington." The telegram was mistakenly included in the dispatch. The publication of the quoted phrase served as

grounds for the arrest. The negligence of the AP in circulating this contraband information throughout the country had no repercussions. J. T. Scharf, *History of Baltimore City and County* (Philadelphia, 1881), pp. 624–625; Andrews, *The North Reports,* pp. 212–213; Harper, *Lincoln and the Press,* 156–158. In another case, the AP circulated a news item from the *Indianapolis Daily State Journal* that had the potential of causing real damage to the army (pp. 331–333). For other incidents of the same nature, see Andrews, *The North Reports,* pp. 96, 113–115, 295, 334, 415, 458–459; on an evasion of censorship by the AP see Gobright, *Recollections,* pp. 316–318, and on leaks, see p. 324. There was one reported case in which an AP agent was arrested for reporting negatively on army morale. Andrews, *The North Reports,* p. 339.

114. Randall and Donald, *Civil War,* p. 467.
115. For a general discussion of legal ramifications of this issue, see Ithiel de Sola Pool, *Technologies of Freedom* (Cambridge, Mass.: Harvard University Press, 1983), especially pp. 75–107.

7. "A Double-Headed Monopoly"

1. "I consider the control now exercised by the telegraph over the press a subject of very great importance . . . The Western Union Telegraph Company and the New York Associated Press [form] a double-headed monopoly." Testimony of Gardiner G. Hubbard, U.S. Congress, Senate, *The Postal Telegraph,* S. Rept. 242, 43d Cong., 1st sess., 1874, p. 105; see also p. 81.
2. An earlier attempt by former members of the Press Association of the CSA in February-March 1866 to reorganize and operate an independent wire service controlling its own local telegraph system failed. John S. Thrasher to John M. Keating, 27 February 1866; 1 March 1866; 3 March 1866; W. G. Clark to Keating, 11 March 1866; Thrasher to Keating, 3, 7, 18 April 1866; 13 June 1866. Finally, the southern press reorganized its association in the spring of 1867 and became a tributary of the NYAP. W. G. Clark, private circular letter, Mobile, 12 March 1867; Clark to Col. [?] 11 March 1867; letters from Thrasher to Keating, March through August 1867. All can be found in the Keating papers, Library of Congress.
3. D. H. Craig to George B. Hicks, NYAP Cleveland agent, 8 December 1862, William Henry Smith papers, Indiana Historical Society, Indianapolis, Indiana; New York Associated Press, *Annual Report* for 1862 (published 1863).
4. New York Associated Press, *Annual Report* for 1862.
5. This was on 21 March 1865, under Michigan law.
6. See, e.g., Horace White to Manton Marble, 17 December 1866, Marble

papers, and see also Western Associated Press, *Proceedings* (1876), p. 34.

7. J. Cutler Andrews, "The Southern Telegraph Company, 1861–1865: A Chapter in the History of Wartime Communication," *Journal of Southern History* 30 (August 1964): 318–344. Thompson, *Wiring a Continent*, pp. 405, 413–415.
8. *New York Herald*, 25 December 1866.
9. This was after the association declined to permit the *Herald*, on 1 August 1866, to publish cable news exclusively. St. Louis *Daily Missouri Democrat*, 8 August 1866, quoted in Peter R. Knights, "The Press Association War of 1866–1867," *Journalism Monographs* 6 (1967), p. 15; see also *New York Herald*, 7 and 11 August 1866.
10. Nathaniel D. Bangs to Marble, 22 and 25 August 1866, Marble papers.
11. The cost was $4,000 in gold. *The Telegrapher*, 1 September 1866.
12. Bangs to Marble, 25 August 1866; William Henry Hurlbert to Marble, August 1866, Marble papers. The best analysis of this development is in "The New York Press Gang," published in the *Chicago Tribune* and reproduced in the *New York World*, 11 December 1866.
13. Hurlbert to Marble, August 1866; Bangs to Marble, 11 and 13 August 1866 (two letters); Hurlbert to Marble, 17 August 1866; Hurlbert to Marble, Saturday [August 1866], Marble papers.
14. "The Associated Press has nobody at the other side to make up its telegrams." Hurlbert to Marble, 17 August 1866, Marble papers; *New York Herald*, 4 and 7 August 1866, and 7 November 1866.
15. D. H. Craig, *To the Members of the New York Associated Press*, printed circular (November 1866); Craig, *To Editors*, printed circular (9 November 1866); Craig, *New York Merchants Exchange and News Association*, printed circular (17 January 1867), Manton Marble papers.
16. Craig to Lefferts, "Thursday," Marshall Lefferts papers, New York Historical Society; Craig, *Testimony*, p. 1269.
17. Craig, *To the Members;* Craig, *To Editors;* Craig, *Testimony*, p. 1269.
18. Craig subsequently wrote Lefferts that, in fact, the initiative came from Western Union: "The quarrel with the Associated Press rose solely from my efforts to comply with [Jeptha H.] Wade's and McAlpin's desire that I should put the Press under the control (as to news business) of the Western Union Co." Craig to Lefferts, Lefferts papers; see also Craig, *Testimony*, p. 1269.
19. Thompson, *Wiring a Continent*, pp. 409–412, 426. Western Union's excess of transmission potential made it possible to promise Craig exclusive wires to all important commercial centers throughout the country. This weakens Knights's proposition that the press association war was concluded because of limited wire capacity.
20. Craig, *To Editors;* Craig, *To the Members;* Craig, *Testimony*, p. 1269; Craig to Lefferts [1871?], Lefferts papers; Craig, *Press News Reports and*

Letter Correspondence by Machine Telegraphy, pamphlet (August 1888), pp. 13, 15. Western Union also agreed to rent rooms and offices for Craig's agency. Western Union, Executive Committee, Minutes, 21 November 1866, Western Union Corporate Archives, Upper Saddle River, New Jersey.

21. Craig provided some details of the premature exposure of his scheme in his circular of 9 November 1866, *To Editors*. See also *New York Times*, 7 and 12 November 1866; *New York World*, 10 November 1866; Craig's card in response, *The Telegrapher*, vol. 3, 15 November 1866.
22. *Chicago Tribune*, 28 and 29 November, 1, 6, and 7 December 1866; WAP, *Proceedings* (1876), pp. 36–50; Knights, "Press Association War," pp. 26–31.
23. The westerners had apparently been negotiating with Craig for some months before Horace White and Murat Halstead arrived in New York (the former on 23 November, the latter on the 25th). WAP, *Proceedings* (1876), p. 41; see also Gramling, *AP*, p. 64.
24. Western Union Telegraph Company, Executive Committee, Minutes, 8 November 1866. Western Union Corporate Archives.
25. For accounts of the meeting see *Chicago Tribune*, 12–17 December 1866; WAP, *Proceedings* (1876), pp. 55–77; Knights, "Press Association War," pp. 42–46. See also Horace White to Marble, 14 December 1866 (telegram) and 17 December 1866, Marble papers.
26. Resolutions of the Executive Committee of the New York Associated Press, 27 November 1866 (manifold copy); Resolutions of the Associated Press, 28 November 1866; Resolution of the New York Associated Press, 4 December 1866; Simonton to Marble, 1 December 1866; Marble to the General News Association of the City of New York, 2 December 1866; George Jones, Samuel Sinclair, and Erastus Brooks to Marble, 7 December 1866, Marble papers. *New York World*, 3, 4, 6, and 7 December 1866.
27. *New York World*, 19 and 21 December 1866, 1 January 1867; *Chicago Tribune*, 20 December 1866.
28. Marble to Medill, 4 January 1867; Marble to Medill, Richard Smith, and Halstead, 8 January 1867; Medill to Marble, 13 January 1867, Marble papers; *Cincinnati Gazette*, 11 January 1867; *Chicago Tribune*, 12 January 1867.
29. Craig, memorandum, Lefferts papers; Norvin Green to Craig, 20 May 1878, Western Union Presidents' Letter Books (hereafter PLB); Western Union, Executive Committee, Minutes, 22 February and 11 March 1867; Joseph Medill to Marble, 21 April 1866, and White to Marble, 23 April 1866, Marble papers; WAP, *Proceedings* (1876), pp. 78–79.
30. Craig to Lefferts, Lefferts papers; Craig, *Press News Reports;* Western Union, Executive Committee, Minutes, 8 November 1866.

31. Lester G. Lindley, "The Constitution Faces Technology: Relationship of the National Government to the Telegraph, 1866–1884" (Ph.D. diss., Rice University, 1971), pp. 41–76; Richard B. DuBoff, "The Rise of Communications Regulation: The Telegraph Industry, 1844–1880," *Journal of Communication* 34 (Summer 1984): 52–65.
32. *Congressional Globe,* 39th Cong., 1st sess., appendix, 382, 24 July 1866.
33. *Congressional Globe,* 39th Cong., 2d sess., 133, 14 December 1866. See also *New York World,* 8 December 1866.
34. Bassnett to Caton, 1860, Caton papers. See also Orton to Weaver, 1 October 1875, PLB.
35. Copies of the contracts appear in WAP, *Proceedings* (1876), pp. 90–94; *Circular from the Executive Committee of the Western Associated Press* (Cincinnati, 15 April 1867).
36. Halstead to Marble, 15 February 1867; White to Marble, 9 (source of quotation) and 23 April 1867, 21 August 1867; Medill to Marble, 21 April 1867; Bangs to Marble, 24 August 1867. All can be found in the Marble papers. Orton to Van Horn, 9 August 1867; Orton to Bennett, 10 August 1867; Orton to Anson Stager, 18 November 1867, PLB.
37. Bennett to the General News Association of the City of New York, 29 June 1867, original in the George Jones papers, and flimsy in the Marble papers; Marble to the General News Association of the City of New York, 2 December 1867 and 3 January 1868, Marble papers.
38. Orton to Simonton, 22 and 28 September 1867; Orton to A. H. Painter, 9 November 1867; Orton to Anson Stager, 18 November and 13 December 1867, PLB.
39. Orton to Simonton, 30 January 1868; Orton to Tinker, 1 February 1869; Orton to Samuel Sinclair and Executive Committee of the NYAP, 4 February 1869, PLB.
40. Orton to Norvin Green, 5 January 1869; Orton to Simonton, 20 January 1869; Orton to Bowles, 25 January 1869; Orton to Tinker, 1 February 1869; Orton to Samuel Sinclair and Executive Committee of the NYAP, 4 and 10 February 1869, PLB; Hudson, *Journalism in America,* p. 488.
41. Orton to Simonton, 11 and 13 February, 1869; Orton to NYAP, 13 February 1869. The original NYAP contract with the American Telegraph Company, which WU was legally required to observe, forbade the telegraph company to buy or sell news. Yet WU let the contract expire and did not renew it. A copy of the original contract is in the Smith papers, Indiana.
42. Orton to Simonton, 14, 15, 21 November 1867; Orton to Stager, 18 November 1867; Orton to W. Scott Smith, 17 December 1867, PLB.
43. W. Scott Smith to W. H. Smith, 16 January and 1 February 1869, Smith papers, Indiana; Orton to Painter, 29 November 1868; Orton to W. C. Bryant and Co., 5–8 February 1869; Orton to Simonton, 21 and 26 No-

vember 1869, PLB; Maury Klein, *The Life and Legend of Jay Gould* (Baltimore: The Johns Hopkins University Press, 1986), pp. 197–205.

44. Orton to Simonton, 15 (three letters) and 18 May 1869; Orton to Joseph B. Howley, 10 February 1869; Orton to Samuel Bowles, 19 February, 28 May, 4 and 17 June 1869; Orton to Eckert, 20 February 1869; Orton to Stager, 26 and 31 May and 8 June 1869, PLB.
45. Orton to Stager, 26 February, 31 May, and 8 June 1869; Orton to Raymond, 7 June 1869, PLB.
46. Testimony of Charles A. Sumner, 23 February 1875, U.S. Congress, House, Committee on the Judiciary, "Investigation of Western Union Telegraph Company," Unpublished Hearings, RG 233, National Archives; Orton to George, 13 and 18 May 1869; Orton to Simonton, 14 May 1869; Orton to Hubbard, 23 November 1869, PLB. Henry George requested Manton Marble's support on political grounds, as the "outrage" on his paper, the only Democratic organ receiving telegraphic news, "will for a long time to come give the dominant party indispensable possession of the most important part of the Pacific newspaper field." George to Marble, 24 April 1869, Marble papers; see also George to Reid, 24 September and 11 October 1872, Reid papers. Hasson's news service was reinforced by John Russell Young, part owner of the *Philadelphia Post* who was expelled from the *Tribune* after stealing AP items and *Herald* specials for his Philadelphia newspaper. See Simonton to NYAP editors, 5 March 1869, Reid papers, correspondence in the Young papers, Library of Congress, and Schwarzlose, *Newsbrokers*, vol. 2, pp. 17, 56–58.
47. *New York Herald*, 26 August 1869, 6 February, 7, 8, and 29 March, 31 August 1870; various unidentified clippings in the Western Union Collection, Electricity Division, Smithsonian Institution; Orton to Richard Smith, 15 July 1869, PLB. The subsequent fate of Craig's initiative is revealed in Craig to Lefferts, 13 February and 1 March 1871, 24 May 1872, and two undated letters; see also 16 December 1874; 4 March 1886, Lefferts papers; Green to Craig, 20 May 1878, PLB; Craig to Smith, 2 May 1889, WU Collection, Smithsonian; undated note in Craig's hand, Delavan Smith papers. Craig would devote some twenty years to the cause of automatic telegraphy, and he would publish numerous pamphlets on the merits of the invention.
48. Orton to C. J. Folger, 19 March, May 1869; Orton to Tinker, 26 July; Orton to Stager, 28 July 1869, PLB.
49. Medill to Marble, 24, 29, and 30 June 1869, and undated note, Marble papers. See also W. H. Smith to Reid, 7 October 1871, Reid papers.
50. Orton to Richard Smith, 15 July 1869, PLB.
51. Orton to Eckert, 20 February 1869; Orton to John Van Horn, 23 March 1870; similarly, see Orton to Howley, 8 February 1869, PLB.

52. Thomas K. McCraw, "Rethinking the Trust Question," in *Regulation in Perspective,* ed. Thomas K. McCraw (Cambridge, Mass.: Harvard University Press, 1981), p. 19.
53. In virtually all the arguments for or against government involvement in the telegraph industry spanning 1866 to 1884, Western Union's Commercial News Department, representing its control over the nation's commercial information, and the company's relationship with the Associated Press, representing Western Union's threatening political influence, figured prominently. See Blondheim, "The News Frontier," pp. 455–469. Quotations from U.S. Congress, House, *Postal Telegraph in the United States,* H. Rept. 114, 41st Cong., 2d sess., 1870, and U.S. Congress, Senate, *Connecting the Telegraph with the Postal Service,* S. Rept. 242, 42d Cong., 3d sess., 1872, pp. 4–5.
54. Orton to Stager, 26 and 31 May 1869, PLB.
55. Orton to Bentley, 27 February 1869, PLB.
56. Orton to Stager, 26 May 1869, PLB.
57. Orton to Simonton, 13 May 1869; Orton to Bowles, 4 and 17 June 1869; Orton to S. F. Wood, 4 June and 1 September 1869; Orton to L. S. H. Roberts, 16 and 22 June 1869; Orton to Joseph Warren, 22 June 1869; Orton to Hawley, 22 February and 24 June 1869, PLB.
58. Orton to Stager, 31 May and 8 June 1869; Orton to Henry J. Raymond, 7 and 17 June 1869, PLB.
59. U.S. Congress, House, *Report from the Committee on the Post Office and Post Roads,* H. Rept. 32, 40th Cong., 3d sess., 1869, p. 12. Compare U.S. Congress, Senate, *Memorial of the Western Union Telegraph Company,* S. Misc. Doc. 86, 42d Cong., 2d sess., 1872, p. 2. The economist David A. Wells appended to his widely circulated study of 1873 (sponsored by Western Union), *The Relation of the Government to the Telegraph* (New York, 1873), hundreds of newspaper clippings that proved the press's overwhelming opposition to government involvement in telegraphy; similarly, Western Union Telegraph Company, *The Proposed Union of the Telegraph and Postal System: Statement of the Western Union Telegraph Company,* pamphlet (Cambridge, 1869), pp. 128–135.
60. Orton to Stager, 21 June 1869, PLB.
61. Orton to Van Horn, 8 July 1871, PLB. See also Bowles to Reid, 13 October 1871, Smith to Reid, 26 December 1873, Reid papers; Simonton to Smith, 10 July 1876, Delavan Smith papers.
62. See Blondheim, "The News Frontier," pp. 448–451.
63. Smith provided a concise autobiographical sketch appended to a letter to Francis F. Browne, 28 October 1887, Smith papers. Good, concise biographical descriptions are Whitelaw Reid's Introduction to Smith's *A Political History of Slavery* (New York: G. P. Putnam's Sons, 1903); and "Death among the Veterans," *The Journalist* 1 (September 1896):

14–15, 17. The most comprehensive secondary biographical treatment of Smith is Edward C. Gray, "The Career of William Henry Smith, Politician-Journalist" (Ph.D. diss., Ohio State University, 1951), which provides coverage of Smith's politics.

64. F. C. Adams, *The Two Smiths* (Washington, 1880), p. 19.
65. Smith replaced George B. Hicks, Craig's confidant in the West. Hicks was suspected of embezzling WAP funds. See Medill to Smith, 16 March 1871, Delavan Smith papers; WAP *Proceedings* (1869), p. 7; (1876), p. 78; and address of Richard Smith at WAP annual meeting [1893?], Delavan Smith papers. For Smith's system of management, see, e.g., Smith to W. L. Davis, Delavan Smith papers; WAP, *Proceedings* (1869), p. 12.
66. WAP, *Proceedings* (1870), p. 8; (1871), p. 6. See also Smith to Reid, 3 February 1875; Jones to Medill, 13 January 1875, Reid papers.
67. WAP, *Proceedings* (1869), pp. 17–21; (1870), pp. 15–17, 19–21; (1871), pp. 10, 28, 33–35; (1873), p. 22.
68. WAP, *Proceedings* (1872), pp. 24–25.
69. "'Vast Facilities' Smith," *Electrical Message,* 15 December 1891.
70. Richard Smith to Smith, 1 February 1872, Smith papers; Smith to Reid, 17 October 1870; Smith to Reid, 3 and 11 February, 1875, Reid papers.
71. WAP, *Proceedings* (1871), p. 18; (1872), p. 30. The policy subsequently proved a serious blunder.
72. Craig to George B. Hicks, printed circular letter, 8 December 1862; *Contract (Confidential) and Remarks by the Committee,* printed circular, 17 September 1864; New York Associated Press Agreement with Chicago Papers, 25 March 1865, Smith papers, Indiana.
73. WAP, *Proceedings* (1876), p. 35.
74. The other regional associations, established in the aftermath of the press association war (except for the Southern Associated Press), had demanded and received the same prerogative from the New York association.
75. Thus, 14 New York dailies circulated 154,000 copies in 1850—an average of 11,000 for each newspaper. By 1860, the average circulation of each daily was 17,000, a 54.5 percent increase over 1850 (18 newspapers circulated 300,000 copies daily). By 1870 average circulation for each daily increased by 35 percent (from 17,000 to 23,000, there being 26 newspapers circulating 590,000 copies daily).
76. Smith to A. P. Swinford, 16 September 1879, Smith papers, Ohio; J. M. Bundy to Cyrus Field, 19 December 1887, Field papers; see also White to Smith, 9 March 1883, Delavan Smith papers.
77. For summaries of the opposition wire services, see Rosewater, *Cooperative News-Gathering,* pp. 147–156; Schwarzlose, *Newsbrokers,* vol. 2, pp. 57–59, 79–80.

78. It was noted above that no public petitions or resolutions were presented to Congress in the interest of controlling Western Union. As early as 1869, however, a petition was submitted to Congress by the National Typographical Union calling for regulation of the Associated Press. The union complained that because the Associated Press restricted its reports, it was "lessening the demand for our labor." The resolution termed the Associated Press "one of the worst monopolies in existence." U.S. Congress, Senate, *Resolutions of the National Typographical Union,* S. Mis. Doc. 13, 41st Cong., 2d sess., 1869; see also National Typographical Union, *Proceedings of the Seventeenth Annual Session* (New York, 1869), pp. 35, 51–52.
79. U.S. Congress, Senate, *Competing Telegraph Lines,* S. Rept. 805, 45th Cong., 3d sess., 1879, pp. ii, 4.
80. Green to Stone, 5 January 1882, PLB.
81. Smith, Memorandum for Mr. Lawson, Smith papers, Indiana.
82. Klein, *Jay Gould,* pp. 277–282.
83. Hueston to Smith, 7 January and 22 April 1882, Smith papers, Indiana; *Chicago Tribune,* 22, 27, 29, and 31 May 1882; Green to Brooks, 13 April 1882; Green to Hueston, 16 May 1882, PLB.
84. Green to Goodsell, 22 March 1882; Green to Van Horn, 13 April 1882; Green to Smith, 30 March 1882, PLB.
85. Smith to his wife, 27 May and 12 June 1882; Smith papers, Indiana.
86. *Chicago Tribune,* 16 and 17 May 1882; Klein, *Jay Gould,* pp. 312–313.
87. R. Smith to Smith, 29 May 1882; Haldeman to Smith, 8 June 1882; Eckert to Smith, 9 June 1882, Smith papers, Indiana. By November 1882 Gould negotiated the sale of the *World* to Halstead or Smith or both, Halstead to Smith, 1, 2, and 26 November 1882, Smith papers. Subsequently he would offer to sell it to a combination of Smith, Eckert, and Hurlbert. Negotiating the deal with Smith ultimately led to the *World*'s sale to Pulitzer, with Smith acting as mediator in the transaction.
88. Gould to Smith, 2 June 1882; Gould and Morgan to Chmn. NYAP, 2 June 1882 (and see quotation from NYAP reply in the Delavan Smith papers); Smith to R. Smith, 13 June 1882, Smith papers, Indiana; Green to Stone, Jones, and Brooks, 26 June 1882; Green to Morgan, 28 June 1882; Green to Brooks, 17 July 1882, PLB.
89. Smith to Gould, 10 June 1882; Smith to R. Smith, 12 June 1882; R. Smith to Smith, 15 June 1882; Smith to R. Smith, 16 June 1882, Smith papers, Indiana; "Confidential Memorandum for General Eckert," Delavan Smith papers.
90. Signed agreement between Green (for WU) and Smith (for WAP), 22 June 1882, Delavan Smith papers; Eckert to Clowry, 26 June 1882, Smith papers, Indiana.

91. Smith to R. Smith, with Hueston to W. H. French enclosed, 1 June 1882; Smith to R. Smith, 6 June 1882; R. Smith to Smith, 9 June 1882, Smith papers, Indiana.
92. Smith to Reid, 8 July 1882, Reid papers; Eckert to Smith, 8 July 1882; Stone to Hurlbert, 8 July 1882; Hurlbert to Stone, 11 July 1882; Stone to Hurlbert, 13 July 1882, Smith papers, Indiana.
93. Green to Brooks, 17 July 1882; Brooks to Green, 18 July 1882, Reid papers; Eckert to Smith, 21 July 1882; Reid to Smith, 5 August 1882; Medill to Smith, 16 September 1882, Smith papers; R. Smith to Smith, 29 September 1882; Hurlbert to Smith, 29 September 1882, Smith papers, Indiana.
94. Smith to Delavan Smith, 18 October 1882, Delavan Smith papers; Eckert to Smith, 7 October 1882, Smith papers, Indiana; Smith to Reid, 25 May 1885, Reid papers.
95. There were earlier serious attempts to negotiate a settlement, yet they were all blocked by members of the ring. Eckert to Smith, 12 and 15 (two letters) August, 7 October 1882; Smith to Eckert, 16 August 1882; W. H. H. [Hurlbert] to Smith, 3 and 10 October 1882, Smith papers, Indiana; Reid to Smith, 10 October 1882, Smith papers; Smith to Delavan Smith, 19 November 1882, Delavan Smith papers; and Minutes of the NYAP, September 1882, in *Resolutions Adopted by the WAP in Convention in Detroit, 1882* (New York, 1891).
96. Green, Memorandum for Mr. Dana, 27 October 1882; Green, Memorandum for Mr. Smith, 3 November 1882, PLB.
97. John W. Strong to Clowry, 27 (2 letters) and 31 July 1882; W. P. Phillips to Smith, 27 September 1882, Smith papers.
98. Sources on the settlement of 1882 are numerous. One of the many good summaries is Smith to Medill, 24 January 1882, Smith papers, Indiana.
99. See, e.g., Hurlbert to Smith, 29 September 1882; French to Smith, 8 November 1882; Medill to Smith, 30 January 1883, Smith papers, Indiana; Reid to Smith, 10 January 1883, Reid papers; Jones to Villard, 10 January 1883, Villard papers.
100. The agreement was buttressed by the purchase of a controlling block of UP stock by members of the joint committee personally, with the certificates held in trust. Sources on the AP-UP trust are abundant. Among the important summaries are Reid to Dana, 30 June 1891, Smith papers; Smith, Report to the Board of Directors of WAP, 1890, Smith papers, Indiana; Statement of the General Manager; United Press, *Report of Committee,* 18 November 1890, Delavan Smith papers.
101. The *New York Mail and Express* received $3,428 in 1882; $7,402 in 1883; $9,502 in 1884; $5,214 in 1885; $9,299 in 1886; and $10,091 in 1887. Memorandum (on *Mail and Express* stationery), 22 December 1887,

Cyrus Field papers. The aggregate of bonuses to NYAP members was $70,744 in 1883; $91,000 in 1884; $68,000 in 1885; and $93,000 in 1886, memorandum, Delavan Smith papers.

102. A list of member newspapers is given in *The Associated Press* (private circular), 1883, William Henry Smith papers, Indiana.

103. A vast body of correspondence and documentation related to these proceedings may be found in the Smith papers, Indiana and Ohio, and in the Delavan Smith papers. For a summary of these developments see Rosewater, *Cooperative News-Gathering,* ch. 17; Schwarzlose, *Newsbrokers,* vol. 2, pp. 149–166.

8. "Talking to All America"

1. Murat Halstead, "The Varieties of Journalism," *Cosmopolitan* 14 (December 1892): 204–205; see also Dill, *Growth of Newspapers,* p. 53. The quotation in the title is from the *New York Times,* 9 September 1859: "The telegraph gives the speaker an audience as wide as the Union. He is talking to all America . . . immediately, and literally with the emphasis of lightening."
2. Halstead, "Varieties of Journalism," p. 204; and see Richard B. Kielbowicz, "News Gathering by Mail in the Age of the Telegraph: Adapting to a New Technology," *Technology and Culture* 28 (January 1987): 26–41.
3. Halstead to Smith, 12 June 1872, Smith papers, Ohio. Halstead held the same opinion as early as 1869. Smith to Whitelaw Reid, 16 November 1869, Reid papers.
4. Prior to the contract of January 1867, the WAP was allowed 1,800 words daily. In calculating the yearly number of words I have added an estimate of excess over the contracted number of words, based on the average excess for the period of 1867–1869. This excess is given in "Statement of the Whole Number of Words in Excess of the Contract, and the Cost of the Same," WAP, *Proceedings* (1869), p. 17.
5. See, e.g., Agreement between the Western Union Telegraph Company and the New York and WAP, 22 December 1882, Cyrus Field papers, New York Public Library. There is extensive correspondence in the William Henry Smith papers and in PLB concerning the inauguration of the leased wire arrangement.
6. Susan R. Brooker-Gross, "Nineteenth-Century News Definitions and Wire-Service Usage," *Journalism Quarterly* 60 (Spring 1983): 25–26. Reliance on copy provided by the wire service was less in the press of large urban centers and decreased in the course of the 1890's (pp. 26–27).

7. W. B. Somerville, *The Newspaper Service of the Western Union Tel. Co.,* printed circular (December 1888).
8. The WAP paid WU $97,567 for its 3,433,000 words of report in 1871, and $116,843 for more than 7,000,000 words in 1881. Throughout the 1870's, the WAP paid WU yearly an average of $106,000 for the transmission of its news. WAP, *Annual Report* (1871–1882).
9. O. Smith to William Henry Smith, 27 December 1873, W. H. Smith papers, Indiana. Detailed maps of southern news circuits (probably of the late 1880's) are in the Delavan Smith papers.
10. In the case of most "pony" circuits, and the service to some of the tributary press associations (for example, the Northwestern PA), WU sold the news directly to the newspapers and paid the Associated Press 10 percent of the collections for use of the news.
11. In 1866 WU operated 66,000 wire-miles; by 1869 it had constructed 18,127 more miles of wires, bringing the total to 106,000. In 1870 it operated 133,000 wire-miles; and in 1880, 291,200. WU Telegraph Company, *Annual Reports,* and Richard B. DuBoff, "The Telegraph and the Structure of Markets in the United States, 1845–1890," *Research in Economic History* 8 (1983): 253–277.
12. Smith to Editor of *Lima Times,* 28 September 1878, Smith papers, Ohio. In 1867, the three Cincinnati members paid $275 for the WAP report, whereas the *Sandusky Register* was charged $25 for the same report. Circular from the Executive Committee of the WAP (Cincinnati, 1867), p. 8.
13. See, e.g., E. B. Gorton to William Henry Smith, 24 May 1871, Smith papers, Indiana.
14. Richard DuBoff found a reduction of as much as one half in the intermarket price dispersion among major eastern markets for agricultural commodities between 1844 and 1854. DuBoff, "Telegraph and the Structure of Markets in the United States." DuBoff used the figures provided by Arthur H. Cole, *Wholesale Commodity Prices in the United States, 1700–1861, Statistical Supplement* (Cambridge, Mass.: Harvard University Press, 1938). See also Cole's own summary of the decline in geographical variation in prices, in *Wholesale Commodity Prices in the United States, 1700–1861* (Cambridge, Mass.: Harvard University Press, 1938), especially table 34 (p. 98). In a study that attempted to isolate the effect of the telegraph and the Atlantic cable on intermarket price differences, Kenneth Garbade and William Silber demonstrated that the new technology brought about the decline of arbitrage virtually to the cost of business transactions. Kenneth D. Garbade and William L. Silber, "Technology, Communication, and the Performance of Financial Markets: 1840–1975," *Journal of Finance* 33 (June 1978): 819–832. See also John Langdale, "Impact of the Telegraph on the Buffalo

Agricultural Commodity Market: 1846–1848," *Professional Geographer* 31 (1979): 165–169; Carey, "Technology and Ideology"; Smythe, "Time, Market, and Space Factors"; Thomas C. Cochran, "The Paradox of American Economic Growth," *Journal of American History* 61 (March 1975): 925–942; Richard B. DuBoff, "The Telegraph in Nineteenth-Century America: Technology and Monopoly," *Comparative Studies in Society and History* 26 (October 1984): 571–586; DuBoff, "Business Demand and the Development of the Telegraph in the United States, 1844–1860," *Business History Review* 56 (Winter 1980): 459–479; and for a general discussion, see George J. Stigler, "The Economics of Information," *Journal of Political Economy* 69 (June 1961): 213–225.

15. By 1870 the western association was able to cut the proportion of the commercial news in its report to about 40 percent. This it accomplished by using a code to condense the market reports. WAP, Report of the General Agent, *Proceedings of the Sixth Annual Meeting* (1870), pp. 10–27; WAP, Report of the General Agent, *Proceedings of the Seventh Annual Meeting* (1871), pp. 12–35; Report of the Board of Directors, *Proceedings* (1871), pp. 6–11.
16. WAP, Report of the Board of Directors, *Proceedings* (1871), p. 7; see also Smith to *Drooer's Journal* [?], 21 February 1873; Smith to J. F. Wallick, 22 February 1873; Smith to Erastus Brooks, 7 December 1882, Smith papers, Ohio.
17. After acquiring control of the Gold and Stock Telegraph Company in 1871, WU merged the operations of the CND with those of Gold and Stock's ticker service in 1873. (Gold and Stock, Minutes of the Meetings of the Stockholders and Directors, 5 January 1874.) In large cities the bulletins received by CND were distributed by various printing telegraphs, see, e.g., "The Quotation Trouble," undated clipping, Norvin Green papers, University of Kentucky. Sending news from the various New York exchanges over the CND circuits helped large operators who bought and sold high volumes on small margins to act with certainty, being informed of hourly price fluctuations in the major markets. Gold and Stock's quotation ticker service was established in 1867. Articles of Association of the Gold and Stock Telegraph Company, 10 August 1867. The Commercial News Department was established by Daniel H. Craig late in 1866.
18. Gold and Stock Telegraph Company, Minutes of the Executive Committee, 5 January 1874; U.S. Congress, Senate, *The Postal Telegraph,* S. Rept. 242, 43d Cong., 1st sess., 1874, pp. 79–82, 94.
19. Gold and Stock Telegraph Company, Report of the President to the Stockholders, Minutes of the Meetings of the Stockholders and Directors, 9 September 1873.

20. For a discussion of this development from the perspective of journalism history, see Leonard, *Power of the Press,* pp. 63–96; from a political science perspective, see William Claggett et al., "Nationalization of the American Electorate," *American Political Science Review* 78 (March 1984): 77–91.
21. *The Press: Its Opportunities and Responsibilities. An Address Delivered before the Members of the Northern Indiana Editorial Association by Richard Smith of the "Cincinnati Gazette,"* pamphlet (Wabash, Ind., 17 June 1881), pp. 9–10.
22. See Daniel H. Craig, "Sworn Statement," *Boston Evening Transcript,* 16 February 1859. Another early statement to the same effect appears in Lawrence Gobright's testimony before a senate committee in 1861. U.S. Congress, House, *Telegraph Censorship,* H. Rept. 64, 37th Cong., 2d sess., 1862, p. 3, and see Poore, "Washington News." Melville E. Stone, *The Associated Press* (Washington, D.C.: The American Historical Society, 1908), pp. 1–4, is a fully developed theory of wire service impartiality by a general manager of the AP. Anticipating these positions, Morse as superintendent of the first government line instructed Vail and Rogers "not to give a partisan character" or "use partisan words" in their news reports. Quoted in *Baltimore and Ohio,* undated pamphlet in the Bates collection.
23. Claims to objectivity by the wire services in general, and the AP in particular, are currently accepted widely by students of American journalism. Indeed, it is to the wire services that many historians give credit for establishing objectivity as a leading professional ideal of American journalism in the second half of the nineteenth century. James W. Carey, "The Communication Revolution and the Professional Communicator," *Sociological Review Monograph* 13 (1969): 23–38; Edwin Emery, *The Press and America* (Englewood Cliffs: Prentice-Hall, 1972), p. 239; see also Schudson, *Discovering the News,* pp. 4–5; Herbert J. Gans, *Deciding What's News* (New York: Vintage Books, 1979), p. 186; Ben H. Bagdakian, *The Information Machines* (New York: Harper and Row, 1971), p. 272; Schiller, *Objectivity and the News,* pp. 2–5. The meaning of "objectivity" is obviously multifaceted and complex. I prefer to focus on the more straightforward notion of conscious and intentional manipulation of information, and on what opposition opinions perceived as bias. But compare Robert A. Hackett, "Decline of a Paradigm? Bias and Objectivity in News Media Studies," *Critical Studies in Mass Communication* 1 (September 1984): 229–259; Tuchman, *Making News,* pp. 156–181.
24. Most of the studies mentioned in the previous note expounding the theory of wire service objectivity are based on the results of a pioneering empirical investigation carried out by Donald L. Shaw, "News

Bias and the Telegraph: A Study of Historical Change," *Journalism Quarterly* 44 (Spring 1967): 3–12; Shaw, "Technology: Freedom for What?" in *Mass Media and the National Experience,* ed. Ronald T. Farrar and John D. Stevens (New York: Harper and Row, 1971), pp. 64–86. Shaw, however, measured bias by counting value statements in news stories. Although in itself an important criterion, it does not reflect bias in news selection.

25. James G. Smart, "Whitelaw Reid: A Biographical Study" (Ph.D. diss., University of Maryland, 1964), pp. 54–86; Donald W. Curl, *Murat Halstead and the "Cincinnati Commercial"* (Boca Raton: University Presses of Florida, 1980); Joseph F. Wall, *Henry Watterson: Reconstructed Rebel* (New York: Oxford University Press, 1956), pp. 99–113.
26. Charles Nordhoff to Worthington C. Ford, 17 April 1884, Charles Nordhoff papers, New York Public Library; *New York World,* 30 October 1885; James G. Smart, "Information Control, Thought Control: Whitelaw Reid and the Nation's News Services," *The Public Historian* 3 (Spring 1981): 23–42. Not much is known on the transfer of the *World* from Scott to Gould, but see, e.g., Julius Grodinsky, *Jay Gould: His Business Career, 1867–1892* (Philadelphia: University of Pennsylvania Press, 1957), pp. 291, 294; Klein, *Jay Gould,* pp. 285, 393–396. Although Gould supported the Republicans, he expressed support for a second term for Cleveland (p. 393).
27. Robert J. Chandler, "The California News-Telegraph Monopoly, 1860–1870," *Southern California Quarterly* 58 (Winter 1976): 459–484. On Simonton's stalwart tendencies see Smith to Whitelaw Reid, 17 May 1872, and Smith to Murat Halstead, 8 November 1879, Smith papers, Ohio. But see also Simonton to Reid, [?] October 1877, Reid papers. Compare Gobright, *Recollections,* pp. 402–407; *One of the Reasons for Telegraphic Reform: Power and Tyranny of the Associated Press—The Character of Its Manager, James W. Simonton,* anonymous pamphlet (1873); Harper, *Lincoln and the Press,* p. 96; see also Ritchie, *Press Gallery,* pp. 51–52.
28. WAP, *Annual Reports,* 1864–1892. The holdings of the Associated Press Corporate Archives, the Associated Press, New York City, and the Indiana Historical Society combined constitute a complete set.
29. On the basis of the WAP annual reports and supplementary biographical data, I surveyed the political affiliation of all WAP officeholders after 1867. Republican domination was evident on the board of directors, though not in the offices of president and vice president. The change in the *Inter-Ocean*'s politics is well documented in the Smith papers. See, e.g., Smith to Rutherford B. Hayes, 22 January 1879; Smith to Sherman, 14 and 30 May 1879, Smith papers.
30. A good illustration of the difference between Watterson and Halde-

man is that while Watterson, on leave from the *Courier Journal* as representative in Congress in 1876, was calling for 100,000 men from Kentucky to converge on Washington and help force the seating of Tilden, Haldeman was supporting the compromise movement both editorially and through negotiations initiated by the WAP. On Haldeman, see Henry Watterson, *"Marse" Henry: An Autobiography* (New York: George H. Doran Company, 1919), pp. 169–176. Haldeman's passivity as executive committee member is best demonstrated in his responses to an interrogation conducted on behalf of the AP of Illinois (1892). Victor Lawson papers, Newberry Library, Chicago. The quotation is from Wall, *Henry Watterson,* p. 151.

31. See, e.g., Charles A. Boynton to Smith, 3 November 1876, Smith papers, Indiana; Boynton to Smith, 3 April 1878 (telegram); 8 August 1878; 26 February 1879; John McCrone to Joseph Medill, 24 January 1873, Smith papers.
32. Quotation from Richard Smith to W. H. Smith, 15 October 1881. Nepotism in the management of the AP brought into the service, as agents, the sons of Richard Smith, William Henry Smith, and I. F. Mack. The three fathers and three sons were Republicans. The data were compiled from various references in William Henry Smith's correspondence.
33. Henry V. Boynton, "The Press and Public Men," *Century Magazine* 42 (October 1891): 853–862. Boynton related the case of a senator who, seeking publicity, made the rounds of the Washington correspondents and promised to furnish each reporter exclusively with a document of great importance. He later distributed the document to each of the correspondents, as well as to the AP. Each newspaper ended up paying some $25 for a special telegram, which duplicated the AP report. "Despatches of the next morning from the home offices to the correspondents in Washington were very emphatic," observed Boynton.
34. Schuyler Colfax to Smith, 28 January and 27 September 1872; Dennison to Smith, 27 January 1872; and Smith to Dennison, 2 February 1872; W. Q. Gresham to Smith, 10 November 1879; James G. Blaine to Smith, 25 June 1877; James Garfield to Smith, 24 February 1881, Smith papers.
35. For other instances of suppression of unfavorable information see Smith to Hayes, 8 April 1878; H. V. Boynton to Smith, 18 February 1878; F. P. Ferris to Smith, 19 and 20 February 1882, Smith papers, Indiana. See also Reid to French, 1 August 1884; French to Reid, 1 August 1884 (telegrams), Reid papers.
36. Governor S. J. Kirkwood to Senator William D. Allison, quoted in Harry Barnard, *Rutherford B. Hayes and His America* (New York: Rus-

sell and Russell, 1954), p. 287. On the publication of letters of support, see Smith to Hayes, 26 January 1876, Smith papers. *New York Times,* 27 January 1876; and see J. W. Simonton to Smith, 8 July 1876 (telegram), Smith papers, Indiana.

37. Kenneth E. Davidson, *The Presidency of Rutherford Birchard Hayes* (Westport: Greenwood Press, 1972), pp. 10–37; Barnard, *Hayes,* pp. 277–294; David S. Muzzey, *James G. Blaine: A Political Idol of Other Days* (New York: Dodd, Mead, and Company, 1934), pp. 83–100.
38. Richard Smith to W. H. Smith, 19 February and 13 April 1876, Delavan Smith papers.
39. H. V. Boynton to Richard Smith, 6 February 1876; Blaine to Richard Smith, 11 February 1876; H. V. Boynton to Richard Smith, 17 February 1876, Smith papers; Boynton to Richard Smith, 18 February 1876, Smith papers, Indiana; Richard Smith to W. H. Smith, 19 February 1876, Delavan Smith papers; W. H. Smith to Richard Smith, 22 February 1876, Smith papers; Boynton to [?] Smith, 25 February 1876; Richard Smith to W. H. Smith, 28 February, 1876, Smith papers, Indiana. See also Keith Ian Polakoff, *The Politics of Inertia* (Baton Rouge: Louisiana State University Press, 1976), pp. 44–52.
40. Blaine to Medill, 29 February 1876 (copy sent to Halstead), Murat Halstead papers, Cincinnati Historical Society.
41. Holloway to Smith, 26 and 28 February 1876, Smith papers, Indiana; Medill to Smith, 29 February 1876, enclosed in Smith to Holloway of the same date, W. R. Holloway papers, Indiana Historical Society; Holloway to Smith, 1 and 13 March 1876, Smith papers, Indiana; Medill to Richard Smith, end of February 1876; Smith to Hayes, 18 April 1876; Smith to H. V. Boynton, 21 April 1876, Smith papers.
42. W. H. Smith to H. V. Boynton, 30 April 1876, Smith papers, Indiana.
43. *Cincinnati Gazette,* 17 April 1876; Smith to Hayes, 18 April 1876.
44. The story was published in the *Indianapolis Sentinel,* a Democratic newspaper. Holloway argued that its publication there was intended to injure Morton by having Republicans suspect, as they did, that the story was leaked by his supporters. Holloway to Smith, 11 and 14 April 1876, Smith papers, Indiana. No one, however, appeared to take that argument seriously. See, e.g., Smith to Boynton, 30 April 1876, Smith papers, Indiana. See also Muzzey, *James G. Blaine,* p. 84. Barnard's account in *Hayes,* pp. 285–286, is somewhat inaccurate.
45. Barnard, *Hayes,* pp. 291–294; Davidson, *The Presidency,* pp. 23–34.
46. Ironically, Hayes, in a letter to Smith, expressed his relief "that no man who is taking an interest in the Hayes movement has had a hand in the affair." Hayes to Smith, 23 April 1876, Smith papers. See also Richard Smith to W. H. Smith, 28 February 1876, Smith papers, Indiana.
47. W. H. Smith to Richard Smith, 22 February 1876, Smith papers.

48. W. H. Smith to Richard Smith, 23 May 1876, Smith papers.
49. W. H. Smith to Richard Smith, 23 February 1876, Smith papers.
50. Hayes to Smith, 2 October 1876, Rutherford B. Hayes papers, Rutherford B. Hayes Memorial Library, Fremont, Ohio; Smith to Hayes, 3 October 1876, Smith papers; V. D. Green to Smith, 24 August 1876; Schurz to Smith, 26 August 1876; L. G. Curtis to Smith, 3 October 1876, Smith papers, Indiana; Smith to Whitelaw Reid, 13 October 1876, Reid papers; Smith to Comly, 16 October 1876, Comly papers, Ohio Historical Society; Gray, "William Henry Smith," pp. 91–93.
51. Smith to Hayes, Smith papers; C. A. Boynton to Smith, 3 November 1876, Smith papers, Indiana. See also Tebbel and Watts, *Press and Presidency,* p. 230.
52. Smith to Hayes, 6 September 1876, Hayes papers.
53. Smith to Hayes, 4 October 1876, Hayes papers; Polakoff, *Politics of Inertia,* pp. 155–157.
54. The newspaper also had close contacts with WU, which cooperated with the AP in tallying the unofficial results, since George Jones, the proprietor of the *Times,* was the NYAP's representative on WU's board of directors.
55. Smith to Hayes, 10 November 1876; Smith to Back [?], (telegram), 9 November 1876, Smith papers.
56. Watterson, *"Marse" Henry,* p. 296.
57. Smith to O. H. Booth, 21 December 1876; Booth's reply, 27 December 1876, Smith papers.
58. *Speech of Charles A. Sumner of California in the House of Representatives, Saturday, March 8, 1884,* pamphlet (Washington, 1884), pp. 28–29.
59. Smith to Hayes, 23 January 1877; see also Medill to Smith, 4 August 1877; Smith to John Sherman, 26 March 1880, Smith papers; Simonton to C. A. Boynton, 25 January 1877; C. A. Boynton to Smith, 22 January 1877; Reid to Smith, 8 February 1877, Smith papers, Indiana.
60. Smith to Reid, 2 February 1877, Reid papers.
61. The Smith papers are the best single source for the compromise of 1876. The standard work on the compromise is C. Vann Woodward, *Reunion and Reaction: The Compromise of 1877 and the End of Reconstruction* (Boston: Little, Brown, and Company, 1951). Woodward also discussed the issue in: Wendell Holmes Stephenson and E. Merton Coulter, eds., *A History of the South,* 10 vols. (Baton Rouge: Louisiana State University Press, 1951), vol. 9, *Origins of the New South,* pp. 23–50. For reappraisals see Allan Peskin, "Was There a Compromise of 1877?" *Journal of American History* 60 (June 1973): 63–79; Michael Les Benedict, "Southern Democrats in the Crisis of 1876–1877: A Reconsideration of *Reunion and Reaction,*" *Journal of Southern History* 46 (No-

vember 1980): 489–524; George C. Rable, "Southern Interests and the Election of 1876: A Reappraisal," *Civil War History* 26 (1980): 347–369.

62. Smith to Hayes, 24 January 1877, Smith papers; see also Boynton, "Press and Public Men," p. 859.
63. Reeves, *Gentleman Boss,* pp. 111–148; Ari Hoogenboom, "Rutherford B. Hayes and Reform of the Spoils System," *Hayes Historical Journal* 4 (Spring 1984): 17–27.
64. Smith to Andrew J. Kellar, 5 May 1878; Kellar to Smith, 8 February 1879; Smith to Hayes, 8 April 1878, Smith papers, Ohio. The course of Smith's investigation is detailed in an extensive correspondence with underlings in the AP and the Chicago custom house and with friends. (Telegrams sent to Smith by his AP agents were charged to the association's telegraph bills, for example, Boynton to French, April 1878, Smith papers.)
65. Smith to H. W. Thompson (telegram), Smith papers.
66. Smith to Carl Schurz, 29 July 1879, Smith papers.
67. This was at the end of the four-year "watershed" period in which Donald Shaw found the greatest decline in value-laden words in political news items in the press. Shaw, "News Bias and the Telegraph."
68. Pulitzer remained largely in Gould's debt for the proceeds of the sale that was arranged and negotiated by William Henry Smith. Correspondence among George J. Gould, William Henry Hurlbert, and Pulitzer, in the Smith papers, Indiana. There is a file relating to the sale in the Delavan Smith papers.
69. Smith to Reid, 16 June 1884, Reid papers.
70. See Reid to Smith, 19 July 1884; 25 July 1884; Reid to French, 1 August 1884; French to Reid, 1 August 1884, Reid papers; Horace White to Smith, 5 September 1884, Delavan Smith papers. On 4 August 1884, Blaine thanked Smith for offering to mediate between himself and Charles Farwell. He stated: "Thank you most sincerely for your constant kindness. Be assured that I profoundly appreciate it." Delavan Smith papers.
71. See Muzzey, *Blaine,* pp. 318–320; Tebbel and Watts, *Press and Presidency,* pp. 264–266; Paul F. Boller, *Presidential Campaigns* (New York: Oxford University Press, 1984), pp. 149–150.
72. French to Smith, 30 October 1884, Smith papers, Indiana.
73. James Simonton, NYAP General Agent, and Thomas T. Eckert, WU General Superintendent, *Instructions for Collection of Election Returns,* printed circular (New York, 10 October 1872), Smith papers, Ohio; *Instructions for Collection of Election Returns,* printed circular (New York, October 1878); Simonton to Reid, 26 October and 2 November 1878, Reid papers.
74. Walter Polk Phillips, United Press General Manager, interview with

New York Herald correspondent, quoted in *The Journalist,* 15 November 1884.

75. *New York World,* 5–7 November 1884; see also WU's statement in reply to the accusations: "Doctoring Returns," 25 November 1884, unidentified clipping in the Norvin Green papers.
76. Norvin Green, President of WU, to N. B. French, President of Commissioners of Police, 6 November 1884, PLB.
77. Green to Pulitzer, 7 November 1884, PLB. What was discussed in the series of meetings behind the closed doors and the rings of policemen is not known. One subject that may have been discussed was Pulitzer's large outstanding debt to Jay Gould. See also Green to Joseph E. McDonald, 20 November 1884, PLB, about the establishment of a committee of inquiry to investigate the report of the Indiana results.
78. David M. Stone, quoted in *New York World,* 10 November 1884.
79. See, e.g., Smith to Reid, 28 May 1885, Reid papers.
80. Smith to Reid, 30 May 1885, Reid papers. Numerous clippings in Reid's scrapbooks demonstrate the large exposure of the charges. See also William Penn Nixon to Smith, 12 November 1884, Delavan Smith papers; and *The Journalist,* 15 November 1884.
81. Representative Charles A. Sumner of California called for an investigation of the role of the AP and WU in the electoral count of 1884. He claimed that James C. Hueston, the man whom William Henry Smith superseded as the head of the NYAP, could testify to "the diabolical scheme for changing the true count of the ballots in the Empire State." *Speech of Hon. Charles A. Sumner of California in the House of Representatives, February 28, 1885,* pamphlet (Washington, 1885), p. 15.
82. Boynton, "The Press and Public Men," p. 858.

Conclusion

1. Reverend J. Lyons, "The Magnetic Telegraph," *New Orleans Daily Picayune,* 6 October 1847.
2. U.S. Congress, House, *Electro-Magnetic Telegraph,* H. Doc. 270, 28th Cong., 1st sess., 6 June 1844, p. 2.
3. *New York Tribune,* 13 November 1847.
4. The statement was made by James Gordon Bennett shortly after the introduction of the telegraph and is quoted in Pray, *Memoirs of Bennett,* p. 365.
5. Quoted in the *Philadelphia Dollar Newspaper,* 17 June 1846.
6. Karl W. Deutsch, *Nationalism and Social Communication* (New York: Wiley, 1953), p. 65. See also Hans Kohn, *American Nationalism* (New York: Collier Books, 1961).

7. Charles Diehl, *The Staff Correspondent* (San Antonio: The Clegg Company, 1931), pp. 37–38.
8. Lewis Mumford, *Technics and Civilization* (New York: Harcourt, Brace and World, 1963), pp. 239–241.
9. For a general discussion of this aspect of political news reporting see Leonard, *Power of the Press*, pp. 70–96. Leonard, following David Donald in *Charles Sumner and the Coming of the Civil War* (New York: Alfred A. Knopf, 1960), argues that the well-known exception to harmonious inter-personal sectional debate—the beating of Charles Sumner in the context of a political discussion—was due precisely to Sumner's appeal to distant audiences in the presentation of his arguments.
10. The course of Davis's sojourn in the North and its repercussions are documented in Dunbar Rowland, ed., *Jefferson Davis, Constitutionalist: His Letters, Papers, and Speeches* (Jackson: Mississippi Department of Archives and History, 1923), vol. 3, pp. 270–354. For summaries see Clement Eaton, *Jefferson Davis* (New York: The Free Press, 1977), pp. 102–106; Hudson Strode, *Jefferson Davis: American Patriot* (New York: Harcourt, Brace, and Co., 1955), pp. 310–313; Robert W. Winston, *High Stakes and Hair Trigger: The Life of Jefferson Davis* (New York: Henry Holt and Co., 1930), pp. 134–136; Robert McElroy, *Jefferson Davis: The Unreal and Real* (New York: Harper and Brothers, 1937), vol. 1, pp. 186–191. See also *Charleston Mercury*, quoted in the *New York Times*, 17 August 1858.
11. The historian David Herbert Donald pointed out this aspect of the breakdown of the two-party system in a lecture on the Civil War at Harvard University, 1986. For background information, see William Nisbet Chambers, *Political Parties in a New Nation: The American Experience, 1776–1809* (New York: Oxford University Press, 1963); Richard Hofstadter, *The Idea of a Party System* (Berkeley: University of California Press, 1972); and Michael F. Holt, *The Political Crisis of the 1850's* (New York: John Wiley and Sons, 1978).
12. *New York Times*, 9 September 1859.
13. Wiebe, *Search for Order*, p. xiii. See also James W. Carey, "The Communication Revolution and the Professional Communicator," *Sociological Review Monograph* 13 (January 1969): 23–38.
14. D. H. Craig, circular printed letter, 1854, Ezra Cornell papers.
15. Shanks, "How We Get Our News."

Acknowledgments

This investigation into the process of accumulating and diffusing knowledge focuses on monopoly: it was made possible, however, by many who generously shared their knowledge with me. Numerous institutions whose function it is to diffuse information did so in a most efficient and amiable manner. These public libraries, historical societies, universities, and archives, and particularly their staff members, were among the most enjoyable discoveries of my research. The encouragement, help, and friendliness I received from William Myers and the staff at the Ohio Historical Society, Eric Pomroy and the staff of the Indiana Historical Society, and many others from Jerusalem to Chicago and from Halifax to Louisville helped make the process of collecting source material the pleasure that it was. I was accorded a kind welcome and assistance by officers of the Western Union Corporation and the Associated Press. Margaret Jasko, Western Union's corporate secretary, did her best to make the weeks I spent at her offices productive and pleasant. John Tomlinson, vice president and general manager of the Associated Press, offered me every help and encouragement.

A Boni fellowship enabled me to study the rich collections of the American Antiquarian Society and to become acquainted with John Hench and his expert staff. Grants from the Charles Warren Center at Harvard University and the Israel Society for Canadian Studies, as well as fellowships from the Lady Davis and Koret Foundations, greatly facilitated my research and writing.

The tradition of intellectual sharing was represented at its finest by the contributions of two scholars. As I began my work, David Herbert Donald helped me clarify my research problems and think through new ways to approach them. He subsequently provided much-needed advice and help in carrying out the research, and finally, he read drafts with a rigor of criticism matched only by the courtesy with which it was offered. An account of his help would be incomplete without also acknowledging the inspiration he provided. Alfred D. Chandler, Jr., gave more of the same. He was always ready with sound advice, useful suggestions, keen criticism, and practical help. I also received valuable aid and advice from Alan Brinkley, Bernard Finn, Samuel Haber, Steve Jaffe, Richard John, Elihu Katz, Richard Polenberg, Stephan Thernstrom, and Martin Van Creveld. With this much help and goodwill from so many, it is a pity that I alone must assume the responsibility for any remaining deficiencies.

My colleagues in the American Studies Department and the History Department of the Hebrew University combined to form an intellectually stimulating and socially ambiant setting for the final stages of the project. Yehoshua Arieli and Aryeh Goren initiated me into the study of American history within that pleasant atmosphere, and with Emily Budick, Shlomo Slonim, and Avi Zakai still help maintain it. I also want to thank the editorial staff at Harvard University Press—in particular Editor-in-Chief Aida D. Donald and Christine Thorsteinsson—for their intellectual input and personal support.

My mother and father, aside from being wonderful parents, were also intellectual companions and excellent critics and editors. Both spent many hours reading drafts and picking up errors in a text written in their, but not my, mother tongue. I wish all three of us could have had more such hours together.

My daughter Roni has lately turned over to her little sister, Shir, major responsibility for tasks such as crayon illumination of Xeroxed documents, the reshuffling of index cards, and making "puter go sleep" before I had managed to save my files. However risky, researching and writing those files in Roni and Shir's company was a pleasure.

My wife, Chagit, didn't erase: nor did she type, make suggestions, criticize, or even read this study. Rather, she made it possible, in too many ways to enumerate. I once read a book by a medievalist who acknowledged his wife for living, many years, "in the shadow of the Crusaders." The pioneers on the American news frontier, at least as exciting a bunch, came to life for me in Chagit's bright light. "*Sheli ve-shelachem shela*" (mine and yours are hers), said the sage of the *Mishnah* Rabbi Akiva to his students, referring to his scholarship and to his wife. I say the same, for Chagit, with love.

Index

Harvard Studies in Business History

1. John Jacob Astor, Business Man, *by Kenneth Wiggins Porter*
2. Jay Cooke, Private Banker, *by Henrietta M. Larson*
3. The Jacksons and the Lees: Two Generations of Massachusetts Merchants, 1765–1844, *by Kenneth Wiggins Porter*
4. The Massachusetts—First National Bank of Boston, 1784–1934, *by N. S. B. Gras*
5. The History of an Advertising Agency: N. W. Ayer & Son at Work, 1869–1949, *revised edition, by Ralph M. Hower*
6. Marketing Life Insurance: Its History in America, *by J. Owen Stalson*
7. History of Macy's of New York, 1858–1919: Chapters in the Evolution of the Department Store, *by Ralph M. Hower*
8. The Whitesmiths of Taunton: A History of Reed & Barton, 1824–1943, *by George Sweet Gibb*
9. Development of Two Bank Groups in the Central Northwest: A Study in Bank Policy and Organization, *by Charles Sterling Popple*
10. The House of Hancock: Business in Boston, 1724–1775, *by W. T. Baxter*
11. Timing a Century: History of the Waltham Watch Company, *by C. W. Moore*
12. Guide to Business History: Materials for the Study of American Business and Suggestions for Their Use, *by Henrietta M. Larson*
13. Pepperell's Progress: History of a Cotton Textile Company, 1844–1945, *by Evelyn H. Knowlton*
14. The House of Baring in American Trade and Finance: English Merchant Bankers at Work, 1763–1861, *by Ralph W. Hidy*
15. The Whitin Machine Works Since 1831: A Textile Machinery Company in an Industrial Village, *by Thomas R. Navin*
16. The Saco-Lowell Shops: Textile Machinery Building in New England, 1813–1949, *by George Sweet Gibb*

17. Broadlooms and Businessmen: A History of the Bigelow-Sanford Carpet Company, 1825–1953, *by John S. Ewing and Nancy P. Norton*
18. Nathan Trotter: Philadelphia Merchant, 1787–1853, *by Elva Tooker*
19. A History of the Massachusetts Hospital Life Insurance Company, *by Gerald T. White*
20. The Charles Ilfeld Company: A Study of the Rise and Decline of Mercantile Capitalism in New Mexico, *by William J. Parish*
21. The Rise and Decline of the Medici Bank, 1397–1494, *by Raymond de Roover*
22. Isaac Hicks: New York Merchant and Quaker, 1767–1820, *by Robert A. Davison*
23. Boston Capitalists and Western Railroads: A Study in the Nineteenth-Century Railroad Investment Process, *by Arthur M. Johnson and Barry E. Supple*
24. Petroleum Pipelines and Public Policy, 1906–1959, *by Arthur M. Johnson*
25. Investment Banking in America: A History, *by Vincent P. Carosso*
26. Merchant Prince of Boston: Colonel T. H. Perkins, 1764–1854, *by Carl Seaburg and Stanley Paterson*
27. The Maturing of Multinational Enterprise: American Business Abroad from 1914 to 1970, *by Mira Wilkins*
28. Financing Anglo-American Trade: The House of Brown, *by Edwin J. Perkins*
29. British Mercantile Houses in Buenos Aires, 1810–1880, *by Vera Blinn Reber*
30. The British Shipbuilding Industry, 1870–1914, *by Sidney Pollard and Paul Robertson*
31. Moving the Masses: The Evolution of Urban Public Transit in New York, Boston, and Philadelphia, 1880–1912, *by Charles W. Cheape*
32. Managerial Hierarchies: Comparative Perspectives on the Rise of Modern Industrial Enterprise, *edited by Alfred D. Chandler, Jr., and Herman Daems*
33. Big Business in China: Sino-Foreign Rivalry in the Cigarette Industry, 1890–1930, *by Sherman Cochran*
34. The Emergency of Multinational Enterprise: American Business Abroad from the Colonial Era to 1914, *by Mira Wilkins*
35. Kikkoman: Company, Clan, and Community, *by W. Mark Fruin*
36. Family Firm to Modern Multinational: Norton Company, a New England Enterprise, *by Charles W. Cheape*
37. Citibank, 1812–1970, *by Harold van B. Cleveland and Thomas F. Huertas*
38. The Morgans: Private International Bankers, 1854–1913, *by Vincent P. Carosso*

39. Business, Banking, and Politics: The Case of British Steel, 1918–1939, *by Steven Tolliday*
40. Enterprising Elite: The Boston Associates and the World They Made, *by Robert F. Dalzell, Jr.*
41. The History of Foreign Investment in the United States to 1914, *by Mira Wilkins*
42. News over the Wires: The Telegraph and the Flow of Public Information in America, 1844–1897, *by Menahem Blondheim*

(Some of these titles may be out of print. Write to Harvard University Press for information and ordering.)